KGB TRAINING MANUAL

COMMUNICATION WITH AGENTS

СВЯЗЬ С АГЕНТУРОЙ

1970

**Translated from the original Russian
by Major Christoph P. Schwanitz (Ret.)**

**Conflict Research Group
London, 2024**

Original edition published for internal use by the Ministry for
State Security (KGB) of the Union of Soviet Socialist Republics,
1970

This English Language translation published by Conflict Research
Group, London, United Kingdom, 2024

About the Conflict Research Group

Conflict Research Group (CRG) is a non-profit think-tank based in the United Kingdom, dedicated to advancing understanding of the art and science of Unconventional Warfare. With a focus on the academic study of guerrilla warfare, revolutionary warfare, asymmetric warfare, Fourth Generation Warfare, Fifth Generation Warfare and political unrest, CRG's work sheds light on the complexities and nuances of modern conflicts. By bringing critical and key works back into print, the organization serves as a vital resource for academics, policymakers, and military professionals seeking in-depth knowledge in these specialized fields.

At the heart of CRG's mission is the belief that a comprehensive understanding of Unconventional Warfare is essential for addressing contemporary security challenges. The group's research and publications delve into historical and contemporary case studies, exploring the strategies, tactics, and implications of irregular warfare. Through this rigorous analysis, CRG contributes to the development of more effective and adaptable strategies for dealing with non-traditional threats.

One of the key aspects of CRG's work is its publishing arm, which is dedicated to bringing into print seminal works on Unconventional Warfare. The group's publications cover a wide range of topics, from historical accounts of guerrilla movements to theoretical analyses of contemporary conflict dynamics and of course reprints of historical official publications. By making these works accessible to a broader audience, CRG aims to enrich the discourse on Unconventional Warfare and contribute to the development of more nuanced and effective approaches to resolving conflicts and disrupting, degrading and defeating unconventional threats.

CRG's research is categorised by its interdisciplinary approach, drawing on insights from military history, political science, sociology, and international relations. This holistic perspective allows the organization to address the multifaceted nature of unconventional warfare, considering not only military tactics, but also the granularity of the political, social, and economic dimensions of conflicts. Through this comprehensive approach, CRG provides a deeper understanding of the root causes and long-term implications of irregular warfare.

Translator's Note

I grew up in East Germany during the 1970s and 80s at the height of the Cold War, with an uncle who was a high-ranking official in the notorious and now thankfully defunct East German Ministerium für Staatssicherheit or Ministry for State Security, more commonly known today as the Stasi.

I did not know it at the time, but before moving into his highly prestigious and important role as Stasi chief for East Berlin, my uncle had spent a large part of his career in the Stasi's Hauptverwaltung Aufklärung section (HVA) which dealt with foreign intelligence operations. This was the GDR's equivalent of the Soviet KGB or the American CIA or the British MI6.

As a young man, particularly during that chaotic period immediately following German Reunification, this all conspired to give me a deep interest in the intricacies of intelligence operations and ensured that I would later seek to forge a career in the field of intelligence. Of course, I would go on to do just that, serving with several different operational units within the German Federal Army until my retirement from the Bundeswher in 2017.

As someone who had spent decades studying Russian language, culture, military and intelligence structures in the course of my work as an officer in the Army, I knew that translating these manuals would require more than just basic linguistic proficiency. It would demand an intimate knowledge of the nuances of each language, as well as a deep understanding of the cultural, historical, technical and operational contexts in which they were written.

One of the more significant challenges I faced, and the one which is least likely to be of any great interest to a reader of this book, was navigating the complexities of Russian grammar and syntax. Unlike German, which is known for its strict rules and conventions, Russian has a more relaxed system that can make it difficult at times to accurately convey meaning. For example, Russian word order often prioritizes grammatical function over semantic content, making it essential to carefully consider the context in which each sentence appears.

Furthermore, Russian relies heavily on prepositions and case endings to convey subtle shades of meaning, whereas German tends to rely more on verb conjugation and adverbial phrases. This meant that I had to be particularly mindful when translating individual words or phrases, as their meanings could shift significantly depending on the surrounding context.

Another significant challenge was capturing some of the more obscure technical jargon used in these manuals. With many of the KGB manuals in the cache dating to the 1960s and 70s, some of the old Soviet terminology has become obsolete and has been replaced by other terms within the Russian Federation intelligence services. The manuals display a wide array of obsolete and specialized terms for various aspects of intelligence operations, from types of agents to counter surveillance techniques to clandestine communications methods.

As someone who is very familiar with German and other NATO partners' intelligence operations and with their jargon and acronyms, I found myself constantly referencing my own knowledge base to ensure that I accurately conveyed the intended meaning of any given passage that included technical operational language. In those few cases where I could not be 100% sure of a technical term's meaning, I simply extrapolated to the best of my ability.

The fact that many of these manuals remain classified in Russia even today speaks volumes about their significance and relevance. It's likely that some are still being used by Russian Federation intelligence services to train new personnel, while others would have no doubt been declared obsolete but remain sensitive due to the nature of their contents.

I have a responsibility to ensure that these manuals are translated accurately and with appropriate sensitivity. It's not just about conveying technical information; it's also about respecting the cultural and the operational context in which they were written. In many ways, translating these KGB tradecraft training manuals was akin to conducting an archaeological excavation into the past. Each sentence or phrase revealed a piece of history that had been hidden away for decades, waiting to be uncovered and shared with the world.

As someone who has spent years studying Russian language and culture as well as evaluating the potential threats which an adversarial Russian Federation might in the future pose to my homeland and our NATO partners, I'm proud to have played a role in making this significant historical material available for public consumption.

I would like to thank "DC" and "CB" from Conflict Research Group for assigning me the delicate but critical task of translating this important material. Having become well informed of the vital work being undertaken by Conflict Research Group, I am honoured to be of service even in this small way.

I would like to thank my beloved wife, Birgitt, for dealing with my many absences and long days spent locked away in my study working on this material and accepting it all with grace and good humour.

I would like to also thank Birgitt for her assistance in helping me translate certain more complex passages from German to English and for proof-reading the final manuscript to correct my abysmal English language grammar. As always, without her, I would be diminished.

Please note that any errors or omissions in these translated pages which may serve to detract from the original Russian language documents are mine and mine alone.

Christoph P. Schwanitz,
Major, KSA (ret.)
Görlitz, 2024

About the KGB

The KGB was the foreign intelligence and domestic security agency of the Soviet Union. It was established on the 13th of March, 1954, soon after the death of Soviet dictator Josef Stalin and it was dissolved with the fall of the Soviet Union on the 3rd of December 1991. The KGB's First Main Directorate was split off and became the Russian Federation's current foreign intelligence service, the FSB.

In addition to its primary responsibilities for foreign intelligence and domestic counterintelligence, during the Soviet era the KGB also had duties such as safeguarding the country's political leadership, overseeing border troops, and carrying out surveillance of the population.

In this book, we are dealing solely with the foreign intelligence aspects of KGB operations, so we shall look at the KGB's foreign intelligence apparatus.

The KGB's First Main Directorate, also known as the First Chief Directorate, was responsible for intelligence operations outside of the Soviet Union.

The directorate was organised into various directorates, including:

Directorate "R" -   Planning and Analysis,
Directorate "S" -   Illegals,
Directorate"T" -   Scientific and Technical Intelligence,
Directorate "K" -   Counter-Intelligence,

Directorate "OT" -	Operational and Technical Services,
Directorate "I" -	Computers Service (would be known as "IT" today),
Directorate "A" -	Active Measures,
Directorate "RT" -	Operations within the USSR

In addition to the administrative directorates listed above, the First Main Directorate had various "Desks" or "Departments" dedicated to operations in various parts of the world or other specialised functions. These were:

1st Department -	North America
2nd Department -	Latin America
3rd Department -	UK, Australia, NZ, Scandinavia, Malta
4th Department -	East Germany, Austria, West Germany
5th Department -	France, Spain, Portugal, Luxembourg, Switzerland, Greece, Italy, Yugoslavia, Albania, Romania
6th Department -	China, Laos, Viet Nam, Cambodia, North Korea, South Korea
7th Department -	Thailand, Indonesia, Malaysia, Singapore, Japan, Philippines
8th Department -	Afghanistan, Turkey, Iran, Israel
9th Department -	English-speaking countries in Africa (South Africa, Rhodesia/ Zimbabwe, Tanzania, Nigeria, etc.)
10th Department -	French-Speaking Countries in Africa
11th Department -	Liaison with other communist countries' intelligence services particularly Cuban and Warsaw Pact nations (was previously known as the "Advisor's Department")
12th Department -	Covers
13th Department -	Covert Communications
14th Department -	Forgeries
15th Department -	Operational files and archives

16th Department -	Signals intelligence
17th Department -	India, Pakistan, Bangladesh, Sri Lanka, Burma, Nepal
18th Department -	Egypt, Syria, Libya, Iraq, Oman, Saudi Arabia, Kuwait, Sudan, Jordan, Morocco, United Arab Emirates/Trucial States
19th Department -	Soviet Expatriates and Emigres
20th Department -	Liaison with 3rd World / newly independent states

It is the KGB's First Main Directorate which was the publisher of these manuals and is most likely that they were produced by staff of the First Main Directorate's *Directorate OT*, which was responsible for Operational and Technical Support functions.

KGB foreign intelligence networks were operated by a KGB Residence or Rezidentura as it is known in phonetic Russian. Please note that in these translations, we refer to the residencies using the equivalent CIA term Station. This is simply to reduce the possibility of confusion and to differentiate between a Residence and a private residence such as those used as safe houses or clandestine postal addresses. Similarly, within the translation, we refer to the KGB Residence's Resident using the CIA term Station Chief.

The KGB Resident or Station Chief was a legal intelligence officer usually operating under diplomatic cover as a "cultural attache" or similar. Diplomatic credentials gave the Resident diplomatic immunity meaning the security forces of the country in which he was operating could never arrest a KGB Resident. At best they could have him expelled from the country like any other diplomat, but this usually had serious diplomatic consequences. Instead, most countries usually worked out fairly quickly who was KGB within their local Soviet embassy and they usually allowed the KGB Resident to operate, but placed him and other Soviet embassy staff under heavy counterintelligence surveillance.

Typically, a KGB Residency was organised into different sections or "lines". Each section had a separate function which supported operations conducted out of any given Residency. These sections could be further categorised into separate functions - Operational and Support.

Operational sections of a KGB Residency were as follows:

Section "EM" -	Intelligence and surveillance of the activities of Soviet Emigres in the host country
Section "KR" -	Counterintelligence and protective security of the Residency
Section "N" -	Support to "illegal" Intelligence Officers in the host country
Section "PR" -	Economic, military, political intelligence on the host country or region as well as active measures such as black propaganda
Section "SK" -	Surveillance and reporting on Soviet diplomatic staff in the host nation.
Section "X" -	Technical intelligence and advanced technology acquisition and transfer.

Support sections of a KGB Residency were as follows:

Section "OT" -	Technical support
Section "RP" -	Signals intelligence
Section "I" -	Information technology

Support staff not assigned to their own specific section included drivers, signals operators, cipher clerks, administrative staff, finance personnel.

Table of Contents

PART III:
NON-PERSONAL COMMUNICATION

PART IV -
COMMUNICATION USING INTERMEDIARIES

Editor's Introduction

The original Russian language manual this English translation is based on was found on the deep web in a cache of scanned older Soviet KGB training materials in a folder on a Russian language .onion site. It is believed that these materials were posted by a dissident many years ago, perhaps even as long ago as 2010 or 2012 based on file metadata. The cache was later posted on the surface web, where to this day scans of the original Russian language documents can still be found through a simple search on any search engine.

Various think-tanks from English-speaking countries had made promises to translate and publish these materials, but despite waiting over five years for them to do so, no apparent progress has been made. With the Russian invasion of Ukraine in February 2022, it appears that translation and publication of the KGB training manuals is no longer a priority for these organizations. As a consequence, and with no clear end to the Ukraine War in sight at the time of writing, we have gone ahead and translated and published the KGB manuals from the cache ourselves.

Please note that we are not the first to publish English language translations of some of these materials. Circa 2020, enterprising persons unknown, in a blatant cash-grab, ran a couple of these documents through some translation software, probably Google, before dumping the resulting unedited text into a book format for publishing on Amazon. We purchased a copy of each of these translations to see whether there would still be a requirement for our professionally translated editions. Sadly, all were largely unreadable, therefore, we pushed ahead with our project.

As Conflict Research Group deals mostly with unconventional warfare, resistance, and inform/influence operations from the perspective of non-state actors, it would seem to the casual reader that espionage training materials from a former nation-state intelligence agency such as the Soviet KGB would fall well outside our remit.

This is simply not the case. During the Cold War, the Soviet Union, its Warsaw Pact satellites and other communist states such as the People's Republic of China and the Democratic People's Republic of Korea invested many billions of dollars in to supporting subversive and revolutionary groups fighting against western interests from Southeast Asia to the Middle east, to Latin America, to Southern Africa. Soviet support for such groups was not limited to weaponry and war materiel, but also included training in communist political theory, revolutionary and guerrilla warfare and of course, in clandestine tradecraft to allow members of a revolutionary or terrorist group to organize, plan and conduct their activities in secret.

Western-trained security forces typically used extremely effective British, French or American counter-intelligence and counter-insurgency methods to detect and destroy insurgent undergrounds or espionage rings at or before their nascent stage, so there was a requirement for guerrilla or terrorist groups sponsored by the Soviet Union to be given the most effective tradecraft training available in the communist world, and that came from the KGB's First Main Directorate.

Two English language resources which closely follow KGB procedures and concepts can be found in the 1980s-era South African Communist Party pamphlet *How to Master Secret Work* and in the 1970s-era document *Security and the Cadre* produced by a Puerto Rican separatist group operating in the US, the Fuerzas Armadas de Liberacion Nacional (FALN). Anyone reading through those two sources and then reading one of these KGB manuals will soon find examples which appear in all three, sometimes almost word-for-word.

Unlike some Western intelligence services such as the CIA, which train personnel in very specific, complex tradecraft techniques and methodology (some involving literal magician's sleight of hand), the KGB instead concentrated on teaching its personnel general concepts. This forced the KGB operative to become highly adaptable and imaginative in putting those concepts into action in the field. This lack of a "toolkit" of relied-upon tactics, techniques and procedures meant no clear patterns were set, making it just that much harder for western counterintelligence services to anticipate the specificities of a KGB intelligence officer or agent's tradecraft in the field.

In closing, I would like to thank Major Chris Schwanitz for his accurate translations of these materials, as well as for standing by for almost 18 months while we decided whether or not to go ahead with this project. I would also like to thank "DC" and her OSINT team for backtracking the circumstances of how the original scanned documents came to be posted online. Finally, I would like to thank you, the reader, for your interest in "charming vintage spycraft" and for your support of CRS by purchasing this book.

CB
London, 2024

PART I

COMMUNICATION -
A CRITICAL PART OF INTELLIGENCE
OPERATIONS

General Principles of Communication in Intelligence Operations

Communication in intelligence means constant interaction between Intelligence Officers and their agent networks, carried out secretly with the help of special means and techniques.

A secret connection is usually established between the central intelligence apparatus (Center) to the "legal" and illegal intelligence Stations of socialist agencies in capitalist countries, within the Stations, Intelligence Officers and intelligence networks and individual agents.

Critical tasks are conducted through secret communication channels in intelligence, for example:

- The Center manages all the work of its units in capitalist countries, supplies them with the necessary material resources and operational equipment;

- In a timely manner, Stations (known as "Residencies" in KGB parlance, but we shall refer to them using the more common CIA term "Stations") in capitalist countries report to the Center intelligence information, reports on work and proposals for its deployment;

- Intelligence officers of "legal" and illegal Stations manage the intelligence work of the agents, receive materials from them and carry out educational work.

These are just the main tasks. In intelligence work, the

range of tasks and individual problems solved with the help of communications is much wider. For example, the Center may need to:

- Urgently transfer to a Station Chief (or through the Station to an agent) the task of obtaining the necessary document.

- Convey instructions to the agent on how to behave when performing the task;

- Warn an illegal intelligence officer or agent about a threatening danger in order to prevent compromise in a particular intelligence unit, etc.

Measures can be taken on all these issues in a timely manner if intelligence has reliable and operational communication with all intelligence units.

The intelligence units of socialist agencies in capitalist countries may obtain a valuable document or useful intelligence information, but these materials will lose their value if, due to lack of or poor communication, they are not delivered to the Center in a timely manner.

Every intelligence officer must understand that the organization of secret, reliable, uninterrupted and operational communication between the central intelligence apparatus and its intelligence units, as well as within units, is a very important part of intelligence work, without which intelligence organisations cannot actually carry out their activities. Therefore, the issue of organizing communications, its continuous improvement and the development of new methods and means of communication at all levels of intelligence is constantly in the center of attention of the entire intelligence apparatus, to include all intelligence officers.

Long experience in intelligence work shows that communications is the most vulnerable point, and that the largest percentage of mission compromises and disruptions in intelligence work occurs due to poor organization or utilization of channels and communication lines.

No matter how secretly and reliably the connection between intelligence officers and agents is organized, the connection between intelligence officers within the station and the connection between the station and the Center depends to a large extent on correct communications procedures. The success of the entire intelligence service depends on it.

In recent years, counterintelligence activities of capitalist countries have intensified against official representatives of socialist states, which has significantly complicated the conditions under which employees of "legal" stations have to maintain contact with agents. This circumstance requires from intelligence officers, and all levels of the intelligence apparatus the most attentive, a most serious and strict attitude to the issues of organizing communications, a decisive fight against the pattern of creating communication channels, a fight against the use of simplified, unreliable means of communication.

Intelligence officers of socialist states must, based on the specific conditions in the host country, create reliable communication channels at all levels of the intelligence apparatus, which would allow them to successfully complete the tasks assigned to the intelligence service of a socialist state. The complexity of the conditions may be a reason that justifies poor communication organization.

Fundamental Requirements for Operational Communication

The nature and critical importance of reconnaissance tasks carried out through communication determine the principles of its construction, the choice of communication means, and the basic requirements placed on it. The main requirements are secrecy, reliability, and efficiency.

Secrecy refers to keeping communication completely secret not only from those not involved in intelligence work, but also from those not directly related to the communication line and the reconnaissance units it serves. It also includes using methods of communication that ensure protection from decryption and interception by the enemy.

Key factors for maintaining communication secrecy include selecting the right communication means, methods, using modern camouflage techniques for intelligence materials, and establishing a well-organized communication structure between intelligence elements.

Intelligence communication should follow a vertical communication principle, from higher levels to lower levels, and vice versa. For example, from the Central command to the field station, from intelligence officers to agents, etc.

With this principle, the so-called horizontal connection between intelligence units is not allowed, that is, a situation where several Stations, intelligence groups, and agents would be connected with each other. Horizontal communication is not allowed in order to ensure the safety of all intelligence cells. If there was a horizontal connection, it would be more difficult to localize a compromise in the event of an enemy intercepting a communication channel or penetrating an agent network.

However, in intelligence work there are times when one agent knows another agent and maintains contact with him. Thus, a recruiting agent always knows the agent he has recruited; a cell leader knows a group of agents whose work he supervises on intelligence orders. Sometimes one agent may be provided with some information about other agents if this is required by the conditions of the mission, for example, when several agents are performing an operational mission jointly.

When organizing communications, cases of cross-talk must not be allowed connection of communication lines. For example, if an intelligence officer maintains contact with an agent through a transfer point, then another intelligence officer in the station should not use the same transfer point to communicate with his agents. In the event of crossing lines of communication with the enemy's counterintelligence, which seeks to paralyze the intelligence activities of a socialist state, open and intercept its communication channels with agents.

The concept of communication secrecy also includes the ability of the intelligence officer and agent to apply to the environment, behavior of surrounding people, the ability not to stand out from this environment with their appearance and behavior. Particularly harmful and dangerous from the point of view of tradecraft is the tendency to become complacent and fall into patterns in organizing communications and its construction without taking into account the intelligence and operational situation. It is imperative to avoid repeatedly using the same meeting places, hiding places, as well as conducting meetings with agents and other operational communications at the same time.

Confidentiality of communications is also ensured by masking, encoding, and encryption of materials sent through all channels, as well as by systematically checking the reliability of existing communications.

The reliability of communications is, first of all, its uninterrupted functioning and resistance to interception by the enemy in any conditions. Reliability of communications is achieved by using the most appropriate means and methods of communication in a specific intelligence-operational situation, developing clear conditions, providing intelligence units with reliable technical means of communication and the ability of intelligence workers to use them perfectly. To ensure reliable communications, intelligence uses only proven and reliable liaison agents, intelligence couriers, agents who keep safe houses and transfer points, and takes care that their cover is convenient for communication between various parts of the intelligence community. Reliability of communications is also ensured by the qualified use of camouflage of intelligence materials (coding, encryption of transmitted information, photography, the use of microphotography, secret writing).

Communication must operate uninterruptedly in case of any complications in the intelligence-operational situation and not depend on any accidents. The reliability of communication and the uninterrupted operation of it are also ensured by the fact that when organizing it, the question of what changes can be made is first thought through.

Possible complications of the situation arise in the use of separate channels and methods of communication with agents. Intelligence officers instruct agents to switch to covert forms of communication using operational equipment, arrange behind-the-scenes meetings and interviews, develop alarm systems, and create behind-the-woods communication lines.

To ensure reliable communications, intelligence sometimes resorts to "duplication." In the most critical areas, ten or more simultaneously operating communication lines are created. The same intelligence material is sent in copies simultaneously over two lines, which provides a greater guarantee of delivery of this material to its destination. This method is usually used when there is no firm confidence in the reliability of the existing communication line and when intelligence materials need to be delivered on time.

Intelligence must always have spare liaison agents, courier agents, and radio stations in case of failure of the existing intelligence equipment. Spare communication lines are not used in current work, but in order to keep them always ready, they are checked from time to time by sending test messages.

Efficiency of communication will make it possible to establish contacts between intelligence organizations and their individual units in the shortest possible time, receive information from agents in a timely manner, and transfer it to the Center in the shortest possible time. For this purpose, reconnaissance uses radio communication as the fastest means of communication wherever possible and develops specially determined sirens to quickly transmit calls to the single from one reconnaissance unit to another, for example, from a reconnaissance officer to the agent.

The above-mentioned requirements for communication in intelligence - secrecy, reliability, and efficiency - are closely interrelated between themselves. For example, a connection cannot be considered reliable if it is not confidential, and the most secret and reliable communication will be useless if there is no efficiency.

To organize communications between intelligence avenues, intelligence in each specific case selects only those methods and means of communication that can ensure the successful conduct of intelligence work in a given intelligence-operational situation.

Excessive clutter and complexity of the communication system is just as harmful as the use of oversimplified, uniform means and generic methods of communication.

PART II

PERSONAL COMMUNICATIONS
WITH AGENTS

General Principles of Personal Communications with Agents

The personal connection between intelligence officers and agents was and still remains one of the main types of communication in intelligence work. Personal communication refers to secret meetings between an intelligence officer and an agent in the process of their intelligence work.

Personal meetings differ favorably from other methods of communication with agents in that they ensure the accuracy and effectiveness of the briefing that the intelligence officer personally conducts with the agent, and allow necessary operational issues to be resolved more quickly on the spot during a conversation. Personal meetings make it possible to exert political and moral influence on an agent and allow you to quickly respond to certain negative aspects in his behavior or mood. During personal meetings, you can train the agent in the techniques of intelligence work, the use of operational and technical means, and secrecy in work; study the personal qualities of agents, their agent capabilities.

Wilderness meetings are especially important when working with newly recruited agents who are not yet sufficiently trained in tradecraft and are not assigned to intelligence. However, personal communication is most vulnerable to enemy counterintelligence.

In recent years, the subversive work of counterintelligence services of bourgeois states against representatives of the countries of the socialist commonwealth has sharply intensified. Their employees, including intelligence officers working in capitalist countries under the cover of official institutions of the socialist state, are under intense surveillance. This circumstance obliges the intelligence Stations of socialist states to pay special attention to the issue of organizing personal meetings with agents in order to ensure secrecy and safety of their conduct. It is advisable to reduce the number of personal meetings between intelligence officers and agents, maintaining this method of communication, mainly for working with newly acquired inexperienced agents for the purpose of conducting educational work or instruction.

Separate personal meetings should be held that require long-term joint stay with the agent (instructions, educational conversation, discussion of issues of expanding the agent's intelligence capabilities, etc.), and meetings for receiving intelligence materials. To receive and transmit materials, it is advisable to organize short-term (instant) meetings and conduct them in such a way that the transmission and receipt of materials are invisible to outsiders.

The essence of a short-term (instant) meeting for the discreet transfer of materials is that those meeting, without giving the impression that they know each other, transfer intelligence materials to one another, using suitable places and conditions for this: a large crowd of people, for example, at escalators stairs in shops and in the metro, in theater stations, on public transport during rush hours, in post offices, narrow passages in various public spaces, etc. The transfer and reception of material during such a meeting is carried out with counter and sometimes parallel movement.

Careful design of the transfer technique is critical to the success of an instant meeting. If an instant meeting is planned to be carried out during oncoming traffic, then it is necessary to determine exactly in which area or zone the transfer will take place, in which hand the material will be located, how it will be transferred, how those meeting should behave in the event of an unforeseen change situation, etc. The time when those meeting must appear in a certain place is appointed with an accuracy of one minute and must strictly go to withstand.

When conducting personal meetings with agents, it is necessary to constantly change the places and times of meetings, while showing ingenuity and resourcefulness, and avoid templates in organizing meetings. If the place for a meeting with an agent is chosen even successfully, but is used by the intelligence officer for a long time, and even at the same hours, it will certainly be discovered by counterintelligence. Any meeting with an agent must be thought out in advance, taking into account the constantly changing situation, and carefully documented. An effective cover story is necessary both in case of detention or arrest, and in case of an unexpected meeting with persons familiar to the agent or intelligence officer.

Selection of Places and Methods for Conducting Personal Meetings

When choosing places and methods for conducting a personal meeting with an agent, it is necessary to take into account the agent-operational situation, the burden of the meeting and its nature (long conversation, the need to take notes, transfer of bulky or small volumes of materials), official and social position, age and gender of the intelligence officer and agent, according to the local norms of appearance and behaviors.

In order to be able to correctly select suitable, from the point of view of tradecraft, places for a meeting with an agent, you need to know the city well: its crowded and sparsely populated areas, the work of public transport, the location of streets, alleys, squares, parks, squares, places suitable for meetings - circulation of entertainment enterprises, museums, libraries, shops, cafes, restaurants, their opening hours. It is necessary to explore the outskirts of the city, places of country walks, etc.

The intelligence officer must ensure that the meeting place is located away from police posts, from points where counterintelligence and police officers and agents are usually concentrated (specially protected facilities, restricted areas, government agencies, foreign embassies and missions, places of residence of foreign diplomats, places of gatherings of criminal elements, etc.) so that the environmental conditions in the area allow both the intelligence officer and the agent to identify possible surveillance of the enemy.

When choosing a meeting place, one should take into account the working methods of local counterintelligence, the qualifications of its employees and the equipment of their modern operational equipment. The peculiarities of life and customs of the local population also influence the choice of meeting place.

The intelligence officer must also take into account the fact that the intelligence and operational situation may suddenly become complicated, for example, some political events may occur in the city (strike, political demonstration), martial law may be introduced, and in certain areas of the city the situation may intensify police activities, etc. In these cases, the place and time specified for the next meeting may turn out to be unsuitable, so the intelligence officer must contact the agent in advance about the time and alternate meeting place.

The choice of place for a meeting also depends on its nature. If the purpose of the meeting is to conduct briefings, educational conversations, analyze the agent's work and discuss his intelligence capabilities, teach him operational techniques, etc., then it is necessary to choose a place for the meeting that would allow for a long conversation and make the necessary notes. For these purposes, it is advisable to use safe houses, and in the absence of such, you can use places for country walks, restaurants and cafes in areas remote from the city center and outside the city, cars, if the security and operational situation in the country allows.

A meeting intended to receive intelligence materials from an agent, return documents to him or to transmit written assignments can be held in museums, at exhibitions, in libraries, in cinemas and theaters, at box offices, in the elevators of large stores. In this case, the size of the transferred materials should be taken into account. Such meetings are, as a rule, short-term (instant) and are organized so that neither external surveillance, if it were discovered by the intelligence officer and agent, nor any of the outsiders would notice the transmission.

To transfer bulky materials, you can use the exchange of externally similar items, invisible to the prying eye - suitcases, bundles of books, packages, magazines and newspapers in trams, subways, buses, trains, reading rooms, in the cloakrooms of gyms, swimming pools, ski stations, at check-out linen to the laundry, that is, wherever the presence of a suitcase or package would not arouse suspicion among the intelligence officer and agent.

When choosing the location and method for a personal meeting with an agent, it is important to consider the official and social status of both the intelligence officer and the agent. An intelligence officer working in a "legal" station should avoid areas where they may be recognized as an employee of a socialist country, and should choose a location where there is minimal chance of being identified as a foreigner.

If the officer lives in a small city and holds a high-profile position, it would be unwise to hold personal meetings with agents in that city, as their interactions could be easily noticed by acquaintances or local counterintelligence. In such cases, it would be better to meet in another city or rural areas less frequented by locals.

Meetings in rural and natural areas often involve activities like hiking, hunting, or sightseeing. The chosen meeting place should allow the agent to maintain their social status without attracting suspicion.

For example, a wealthy agent dining in a cheap restaurant in a working-class neighborhood would look out of place, while a lower-ranking individual dining in an expensive restaurant could raise suspicions. Meetings with high-profile agents should be conducted discreetly to avoid drawing attention.

Intelligence officers from "legal" stations can utilize diplomatic receptions, official visits, interviews or other legal and seemingly routine opportunities to meet with agents who due to the nature of their employment or other activities have an opportunity to innocently come into contact with representatives of socialist governments. Meetings can also take place in embassy or trade mission premises, but precautions should be taken to prevent eavesdropping. It is important to create cover stories for agents' prolonged absences. In intelligence practice, there were cases when, for example, insufficiently explained long absences of husbands caused jealousy of wives. Thus, the wife of one of the agents, suspecting her husband of adultery, decided to check where he was. She enlisted the help of a private detective, who recorded her husband's meeting with the intelligence officer.

Be sure to consider gender-specific locations in some countries. For instance, some places may be designated for men or women only. For example, in many countries there are bars and restaurants with sections only for men and only for women.

Special Features of Conducting Meetings Under Various Conditions

When choosing a place for a meeting with an agent, the intelligence officer must first, and always personally, examine this place to make sure that it meets the requirements of secrecy, study the routes to the meeting place, determine where it is most convenient to detect external surveillance, define alternative escape routes in case of emergency, and establish communication protocols in case of unforeseen circumstances.

Additionally, the intelligence officer should consider the noise level, lighting conditions, and overall security of the meeting location to ensure the safety and confidentiality of the meeting.

It is crucial for the intelligence officer to be prepared for any potential risks or threats that may arise during the meeting and to have contingency plans in place to handle them effectively.

Let the agent know what type of transport is more expedient, from the point of view of secrecy, to use when going to the meeting place, what time of day is most convenient for using the chosen place.

Personal meetings with agents can be held on the street, in restaurants and cafes, outside the city, in safe houses and in moving cars. Let us briefly consider the features of holding personal meetings in the indicated places.

Meetings on the street.

The positive thing about meetings on the street is that if precautions are taken, participants are to a certain extent guaranteed against eavesdropping and have great opportunities to protect themselves from external surveillance by the enemy.

For long conversations, it is recommended to meet not on noisy, crowded streets, not on large city highways, but on the streets adjacent to them, or in quiet residential areas remote from the city center, in which there are usually no government agencies or industrial enterprises. These streets are relatively sparsely populated, but not deserted; they provide the opportunity for an intelligence officer and an agent to have a conversation in a more secure, calm atmosphere, outwardly no different from the people living in this area who are supposedly returning home from work or, one MAY assume, on the contrary, are heading from home to the city.

When scheduling a meeting outside, you must take into account the possibility of weather changes and provide an alternate sheltered place near the designated meeting location in case of rain.

Meetings in restaurants and cafes.

In most of the countries of Western Europe, in America, large cities in southern Africa and the Middle East, restaurants and cafes have become part of everyday life for almost all segments of the population.

The cafe can provide breakfast, lunch, snacks and coffee or tea, which provides time and an excuse for patrons to loiter in the cafe. A patron can spend unlimited time there reading books, newspapers, playing chess or backgammon, in conversations with friends.

Cafes are visited also for business meetings and conversations. Therefore, it's not difficult for an intelligence officer and agent to create a plausible cover story about being together in a cafe or restaurant.

However, when arranging a meeting with an agent in a cafe or restaurant, you should remember that such establishments are often under constant surveillance by police and counterintelligence agencies. Almost all large cafes and restaurants have police and counterintelligence informants. Therefore, meetings with agents in cafes and restaurants can only be carried out under the obligatory condition that the intelligence officer and agent are not known either to the administration or to the service personnel of the cafe or restaurant, that is, no one from the environment should know ether the cover or clandestine nature of the intelligence officer's or the agent's work.

When choosing a cafe or restaurant, you should check whether they are a gathering place for gamblers, smugglers and other petty or organised criminal elements, since in such cafes police raids and associated document checks are possible, and this can lead to unpleasant consequences for the intelligence agent.

It is not recommended to meet in specialized cafes, visited mostly by foreign tourists, expatriates, diplomats, military personnel, etc. since such cafes are more actively monitored by counterintelligence and police agencies.

The most convenient cafes and restaurants for conducting meetings between intelligence officers and agents are those located near the main roads and in particular, near intersections which allow multiple potential escape routes.

One advantage to using large, popular cafes and restaurants is that such establishments usually have larger parking lots which make it possible to transfer materials even of considerable size.

Meetings outside the city.

Depending on local conditions such as the time of year and the climate, meetings outside the city can be held under the guise of hiking, swimming, fishing, hunting, picnic, skiing. Such meetings are usually well-planned with agents.

Meetings outside the city guarantee to a certain extent, protection from eavesdropping by the enemy and create favorable conditions for identifying external hostile surveillance When meeting outside the city, chance encounters with acquaintances or colleagues of the intelligence officer or the agent are less likely.

Meetings at safe houses.

If an agent, due to his official social position, is well known to the local population, then holding meetings with him in public places or on the streets is extremely undesirable and dangerous. It is not always possible to arrange a meeting with such an agent outside the city. It is recommended to meet with agents from this category of people, especially for long conversations, at safe houses.

Safe houses are used for meetings and when an agent or intelligence officer needs to work on materials, write reports, etc. or when there is a need to train an agent to use certain types of operational equipment, for example: to use radio equipment, a camera, special ink for secret paper, a code, a cipher, etc.

Darkrooms are sometimes set up at safe

houses for photographing intelligence materials that need to be urgently returned to the agent. It is usually disguised as an amateur photographer's laboratory.

When using a safe house, the agent-keeper of the safe house should not be familiar with the essence of the intelligence officer's work, see in person the agents coming to the apartment, and, in any case, should not know their names and other identifying information.

A safe house can be a separate house, a country house, or an apartment in a residential building. To cover up visits to an apartment by agents and intelligence officers, it is advisable to select agents from among doctors, lawyers, singing and music teachers, and foreign language teachers as the owners of safe houses so that strange languages, tailors, etc. do not immediately arouse suspicion and that visits are justified by the profession of the agent of the owner of the safe house.

It is recommended to choose safe apartments in buildings where there are no gatekeepers or elevator operators, and where visitors can enter and exit unnoticed by strangers.

From time to time you should check to see if your apartment or house is being monitored. It is necessary to arrange a danger signal with the owner of the safe house in case the apartment is being monitored by local counterintelligence or if an ambush is set up in the apartment.

When using safe houses for meetings with agents, it should be taken into account that counterintelligence services of capitalist countries, if they have any suspicions about a safe house, can easily organize surveillance of it with the help of neighbors and, which is very difficult to detect. They may also furnish the apartment with eavesdropping and photographing equipment.

In the event of a police raid on a safe house, it is much more difficult than during meetings on the street, in a cafe, to provide a plausible explanation for the joint presence of an intelligence officer and an agent there. Therefore, it is necessary that the visit to the apartment can be justified by the profession of the owner of the safe house.

When visiting a safe house, you should take the necessary measures to detect external surveillance and comply with other secret requirements. For example, the use of the same safe house by several intelligence agents is fraught with danger: the betrayal or compromise of one agent can lead to the compromise of a whole intelligence network of spies using this apartment and their agents.

Meetings at safe houses are recommended to be held only with trusted, valuable agents and in those cases, when there is really a need for it. To conduct meetings with particularly valuable agents, you should have a separate safe house for each agent.

Meetings in a moving car.

When conducting meetings in a car, it is advisable that the intelligence officer or the agent himself drive the car. However, you can meet in a car even when the car is driven by an operational driver. For the intelligence officer, this is even more convenient, because in this case neither he nor the agent have to concentrate on controlling the car and can talk freely. However, it should be borne in mind that the presence of a third party at a meeting usually makes the agent nervous, so it is necessary to prepare him accordingly and convince him that the operational driver is an absolutely reliable person and that he does not know foreign languages.

The question of holding meetings with agents in cars, choosing an agent with whom it is possible and advisable to hold a meeting in a car, the car that can be used for this purpose is decided on depending on the intelligence and operational situation in the country being intelligence targeted, the capabilities of the Station, the relationship between the intelligence officer and the agent and other specific conditions.

Usually the car drives up to a predetermined place where an agent should wait for the intelligence officer. When the intelligence officer and agent are in the car, you should not drive on streets with heavy traffic, since as a result of an accidental violation of traffic rules or due to the fact that the car belongs to an official employee of a representative office of a socialist state, the attention of the police may be drawn to it. The car can be taken under surveillance by police cars and counterintelligence vehicles patrolling in the city, since the license plates of the representatives of socialist states are well known to them.

The route taken by car must be determined ahead of time. A conversation with an agent can be carried out not only during movement of the car, but also at a stop, in any convenient place hidden from the gaze of strangers. After the end of the conversation, the agent must be taken by car to a place where his movement from the car would not have aroused suspicion.

The car used for meetings must be in good condition, have reliable tubes and tires, and must always be filled with a sufficient amount of fuel.

If meetings are held in a car belonging to an intelligence officer, then for camouflage purposes it is desirable that this car be used by the intelligence officer not only for meetings with agents, but also for personal purposes, and, if possible, have not a diplomatic, but a regular city license plate. Covert vehicles of military stations and vehicles used for surveillance or clandestine purposes should not stand out (by make, model and color) from the mass of vehicles operated in the reconnaissance country. Such vehicles would be easily lost among them.

Before meeting with an agent, the intelligence officer must carefully check whether his vehicle is under surveillance. Counterintelligence of some capitalist States resort to secretly installing special equipment in the cars they are monitoring - self-recording equipment and automatic radio transmitters. In this regard, the operational vehicles of the stations must be periodically carefully matted, especially before meetings with agents, regardless of the fact that the vehicles are located in the garages of government foreign service institutions of the socialist state (embassy, trade mission), since the enemy's counterintelligence may find an opportunity to establish develop the specified operational technique. It also follows from this that intelligence officers must refrain from frequent use of cars for meetings with agents, and beware of writing names, addresses and other identifying information on agents and other highly secret data on the road atlas.

In some countries where the car rental system is widely developed, an agent, an illegal intelligence officer or a figurehead hired for this purpose can rent cars from private companies. These vehicles can be used for operational purposes with less fear of attracting the attention of counterintelligence officers of the enemy security forces. When using vehicles for reconnaissance work, the necessary secrecy must be observed.

The following are places where it is not recommended to hold meetings with agents:

a) In the places of residence of agents and official acquaintances of the agent and the intelligence officer, where one cannot be guaranteed against a chance meeting with them by the intelligence officer or agent.

b) in areas where military installations and government agencies are located; in places where prominent government officials live, near specially protected factories, laboratories, research centers, etc.

c) In areas where embassies, consulates and other official foreign institutions are located, as well as in places of residence of representatives of socialist states. These areas are usually under heavy counterintelligence surveillance police.

d) Near large banks, jewelry stores. These areas, especially in the evening, are protected by police and private detective agencies.

e) In areas that have a dubious reputation due to the concentration of criminal elements there and the presence of brothels. Raids are often carried out in such areas, they are also under the supervision of immigration services, special authorities for combating desertion, drug trafficking, etc.

f) In hotels, since they widely use eavesdropping techniques and have a wide network of counterintelligence agents and informants.

Selection of Time for Conducting Meetings

In order to ensure confidentiality of personal meetings, agents should pay attention not only to the choice of the meeting place, but also to determining the time the meeting is to be conducted. The time for the meeting should be chosen taking into account the official and public employment of the agent and intelligence officer.

Usually the preference is to choose evening time for meeting with agents, since at this time it is easier to escape enemy surveillance, it is easier to find secluded deserted places, and it is also easier to create a cover story for the reasons for going out into the city at off-duty hours. However, holding meetings in the evening is not always possible, and sometimes it's even impractical.

Sometimes an intelligence officer is forced to meet with an agent during the day in order to urgently receive from him or return to him documents that can, for example, be handed over only during a lunch break. The need for a meeting during the day may also arise when an agent comes to a meeting from another city and cannot stay until the evening at the intelligence officer's location. Sometimes in the evening the agent is busy with work or cannot attend a meeting in the evening due to family and other circumstances.

The timing also depends on the location of the meeting. For example, it is advisable to schedule meetings in cafes and restaurants at the time of the greatest influx of visitors, since at this time the service personnel are busy and have less opportunity to engage in observation.

When choosing a meeting time, it is also necessary to take into account the rhythm of life of the city and the meeting area; you need to know, in particular, at what time life usually stops on the streets and in public places that are scheduled for the meeting. It is not recommended to make appointments during hours when all traffic on the city streets stops, since the appearance of passers-by on the streets at unusual times attracts the attention of the police officers on duty and night watchmen.

Taking into account local characteristics is absolutely necessary. For example, in some Eastern countries in winter, residents go to bed very early and after six o'clock in the evening there is no one on the streets.

You should arrive at meetings exactly at the appointed time so that neither the intelligence officer nor the agent waits for each other and thereby attracts the attention of others.

In order to disorient local counterintelligence agencies, it is necessary to stagger at random the time for meetings with agents as much as possible, using not only evening, but also daytime and even morning hours, taking into account, of course, the agent's work schedule.

Intelligence Officers' and agents' watches must always show accurate time, and preferably be synchronized to within a few seconds at most. Typically this means synchronizing watches to the local time as accurately as possible.

Preparing Intelligence Officers for Personal Meetings

In modern conditions, when the main counterintelligence forces of capitalist states are concentrated on working against the intelligence services of the countries of the socialist commonwealth, holding a meeting with an agent is a complex and critical matter that requires careful preparation from the intelligence officer. Some meetings involving the receipt or return of classified materials may sometimes involve several intelligence officers.

A few days before the meeting, the intelligence officer draws up a plan for its conduct, in particular, develops a check route, outlines a plan for a conversation with the agent, a new task for him, the next and backup meetings.

When conducting complex meetings, the plan will detail the intelligence officers who will take part in them, the specific tasks of each of them (conducting counter-surveillance, receiving materials from the intelligence officer conducting the meeting and delivering them to the station, photographing materials, etc.) Vehicles and other technical means are indicated, a reserve of intelligence officers and technical means is planned in case the main participants in the event, due to enhanced counterintelligence surveillance or for some other reason, do not have the opportunity to take part in the planned event.

Meeting plans are usually discussed with the Station Chief of the local station or his deputy, and only after approval can the intelligence officer start implementing them. In special cases, the plan for conducting a meeting with the agent is reviewed and approved by the central intelligence apparatus itself.

Before going to a meeting, the intelligence officer must make sure that he is not being surveilled externally by hostile security forces. Any field intelligence officer should be dealing with this issue on a daily basis, and not just on the day of going to a meeting. The check of the route is carried out skillfully so that the enemy's external surveillance service doesn't notice that the intelligence officer is being checked. By all his behavior, the intelligence officer must not give the enemy's counterintelligence a reason to suspect him of intelligence activities or belonging to intelligence. More active external surveillance may be established, which will make intelligence work extremely difficult.

An intelligence officer's excursions into the city must be conducted under reasonable and plausible cover. Use plausible excuses, for example, visiting city parks, cincmas, shops, museums, libraries, theaters, walks around the city, etc. In this case, the behavior of the intelligence officer should not differ from the behavior of other employees of the cover agency. The intelligence officer's regular trips to the city on personal business on days free from meetings, establishing and maintaining contacts with neutral connections - all this disorients the enemy's counterintelligence and very well disguises the intelligence officer's trips to meetings with agents and other intelligence activities. Regular outings will allow the intelligence officer to acquire and improve skills in order to detect external surveillance, develop his ability to identify suspicious people in crowds, practice observation, and increase sharp vigilance.

If an intelligence officer goes out into the city on certain days and only to meet with agents, this will certainly attract the attention of counterintelligence and will make things much easier for the external surveillance service.

If, as a result of a regular preliminary check, the intelligence officer determines that he is under constant surveillance, he must temporarily stop meeting with the agents, without, however, changing his lifestyle, that is, continue to travel to the city on official or personal business. At the same time, the intelligence officer, together with the resident station chief, is trying to find out the possible reasons for the increased surveillance by local counterintelligence in order to take other precautions.

Only after, as a result of a lengthy systematic check, it is established that external surveillance has been removed, the intelligence officer can resume undercover work, continuing, however, to carefully check whether external surveillance has been established again. An exception can only be made under special circumstances, for example, when there is a need to hold an urgent meeting with a courier or return materials to an agent.

In such cases, the intelligence officer must think over the route more carefully and take measures in advance to ensure that his exit from the house or institution remains unnoticed by the intelligence officers of the external surveillance service.

If an intelligence officer on a surveillance detection route discovers that he is being observed, he must escape from it, using pre-designated places for escape.

In some cases, in order to carry out important intelligence activities, the station takes measures to cut up the counterintelligence forces. In particular, before an intelligence officer conducting one or another operational activity enters the city, other intelligence officers are organized to enter the city, having no missions other than to tie up and divert the attention of the external surveillance operatives. The intelligence officer conducting the meeting must leave the mission unnoticed. Sometimes a covert operational vehicle is used for this. An operational driver or another intelligence officer, unnoticed by counterintelligence, covertly takes the intelligence officer conducting the meeting to the city. After a thorough check, the vehicle stops in a convenient place, the intelligence officer exits, departs from it and continues to follow the developed route to the meeting point, undertaking additional checks on foot and by public transport.

If the meeting is to take place during the day, during working hours, then a visit to the cinema or other entertainment enterprises should not be chosen as an excuse for going out into the city, since this may arouse suspicion. During official hours, the justification for an intelligence officer to go out or go to a meeting with an agent can be business visits, a visit to a doctor, a tailor, etc.

It is advisable to document or "backstop" the cover for trips to meetings in remote places and the associated absence from cover service as business trips or short-term trips from deployment (1-2 days) for the purpose of visiting some sights in another city. Such cover is also necessary for local authorities, since in many capitalist countries, trips around the country for employees of representative offices of socialist states are associated with the need to warn the Ministry of Foreign Affairs about the trip in advance and obtain special permission from local authorities.

If you need to travel to a meeting by train, then special care must be taken when purchasing a ticket. In cases where tickets are pre-sold, it is sometimes better to entrust the purchase of the ticket to another intelligence officer who knows the local language well, or even to an agent when the nature of the trip justifies it. In most cases, intelligence officers purchase tickets for themselves with a full guarantee that counterintelligence will not record their departure from the city.

If the city where the meeting is scheduled is located close to the location of the Station, it is more advisable to purchase a free reserved seat ticket, which allows you to travel in any carriage. A free-of-charge ticket is convenient for an intelligence officer because it is valid on any train, and in a number of countries - for several days. In the USA and England, as well as in some other countries, tickets (except for reserved seats) can be purchased on the train from the conductors.

Sometimes it is advisable to buy a ticket only to an intermediate city in order to do a short check again, and then go to the city where the stopover will take place. You can practice boarding the train at the first station closest to the city where the intelligence officer is to stay, which can be reached by car, having checked appropriately along the route.

If at all possible, it is recommended, in order to check the presence of external surveillance, to get off the train before reaching the city planned for the meeting, or having passed it, in order to check in some neighboring settlement, arrive at the meeting point by local train or other means of transport.

Departure for a meeting must be done in advance so that the intelligence officer has the necessary time to complete scheduled (for the purpose of camouflage) official or personal matters.

Knowing the city well, an intelligence officer can successfully conduct a check and detect surveillance if it is being observed. If surveillance is detected, the intelligence officer, as a rule, does not go to the meeting and does not approach the meeting place. Only in the most extreme cases, for example, when it is necessary to return time-sensitive materials received from him to the agent, does the intelligence officer take measures to avoid external surveillance and, after thoroughly checking, go to the meeting place with the agent. But even in these cases, the intelligence officer must always strive to make his escape from surveillance look natural, and not deliberate, so that the employees of the external surveillance service do not suspect that they have been compromised, but instead get the impression that they themselves have lost him.

It is necessary to carefully check not only on the way to the meeting place, but also during the meeting and after leaving the meeting place, strictly observing the counter-surveillance rules.

If an intelligence officer discovers surveillance during a meeting, he should pretend that he did not notice it. You should shorten the conversation, but end the meeting naturally, as usually happens when acquaintances meet, without ending the conversation abruptly. The intelligence officer must alert the agent to the compromise. This should be done very carefully so as not to scare the agent. The intelligence officer may, for example, say that he does not like the behavior of such and such a person, that, just in case, the agent should carefully check after the meeting to make sure that he is not under surveillance, and if hostile surveillance is confirmed, take action to get away from it. The intelligence officer should also quickly check that the agent remembers the meeting cover story well.

If, at the end of a meeting held in another city, an intelligence officer discovers that he is under surveillance, he must, based on the fact that the enemy surveillance operatives may not know who he is, take action and skillfully evade surveillance.

An exception is allowed only when the intelligence officer and the agent officially know each other and their meeting can be "legalized". In this case, it is not always necessary to avoid observation, as this may raise suspicion that the meeting took place in secret. This also applies when an intelligence officer holds a meeting with an agent in the city where the station is located, where he is known to counterintelligence by official status.

It is strictly forbidden to follow from one meeting to another, even if the intelligence officer did not notice External Surveillance, since in the presence of external surveillance, which the intelligence officer could not detect, not only the first, but also the second agent may be under surveillance.

The intelligence officer must accustom the agent to advance warnings when going to meetings, teach him to check himself for surveillance along the way, determine with him in advance the line of his behavior if he notices he is under external surveillance. In particular, if he discovers that he is being watched while still on his way to the meeting place, he, like the intelligence officer, should not make attempts to escape from the surveillance and should not proceed to the meeting place. If an observation is detected by an intelligence officer and an agent directly at the meeting place, a special alarm is developed, with the help of which those meeting, without approaching each other, give a danger signal.

Warning signals must be such that they do not attract attention and are not noticed by enemy surveillance personnel. Such a signal can be, for example, a book, newspaper, gloves, cigarette, etc., located in a specified position. If an agent is waiting for an intelligence officer in a cafe or restaurant, then such items can, for example, be placed on a table. You can signal danger by some movement, for example, taking off your hat, lighting a cigarette, rubbing your eyes, etc. But such signals must be used carefully, since enemy surveillance operatives will pay attention to them. They will try to determine who the signal might have been given to if it was given unnaturally.

The intelligence officer should not completely rely on the agent to detect hostile surveillance; he himself should periodically check whether the agent is under surveillance.

For these purposes, you can resort to the following method: two meeting places are agreed upon with the agent. In the first place, the intelligence officer and the agent establish only eye contact with each other. After this, the agent goes to the second place where the meeting itself should take place. On this route, the intelligence officer conducts counter-surveillance drills to check whether the agent is under surveillance. If an observation is detected, the intelligence officer does not go to the meeting place, which is self-evident. This in itself is a danger signal for the agent. The places of these meetings should be chosen in such a way that the route from one place to another allows the intelligence officer to carry out his surveillance checks.

The initiative to establish contact in all cases must come from the intelligence officer. Before receiving the ready to meet signal, the agent should not approach him and pretend that he knows him.

If the meeting is held on the street or in any public place, it should look like an ordinary business or friendly meeting. An intelligence officer and an agent should not whisper, pretend to be mysterious, break off the conversation mid-sentence when strangers approach them, or look around often.

The behavior of the intelligence officer and agent must also be appropriate to the meeting place. If a meeting is scheduled, for example, on a tennis court, then those meeting need to play tennis and then hold a conversation under the guise of relaxation.

While waiting at the meeting place, neither the agent nor the intelligence officer should show by their behavior that they are looking forward to meeting or that they are waiting for something.

If the intelligence officer does not speak the local language perfectly, then in a cafe, restaurant and other public places, in order to avoid deciphering his foreign origin, he should not talk to the service personnel, leaving this to the agent. In this regard, the intelligence officer should not arrive at a meeting in such places before the agent.

Preventive Measures in Case of Arrest of Intelligence Officer and Agent

When conducting a meeting with an agent, one should always consider the likelihood of the intelligence officer and agent being detained by enemy security forces or counterintelligence agencies.

Arrest can follow not only as a result of the preliminary development of an agent or intelligence officer, but also by chance, for example, during a raid, a check of documents by a police or military patrol.

The intelligence officer must take precautions in advance. Every meeting between an intelligence officer and an agent must be conducted under a believable cover story. It is necessary to create an effective cover story to explain the circumstances of the meeting, the reasons for the meeting, and the nature of the issues discussed.

The cover story is developed in advance so that the intelligence agents understand it well and so that their testimony in the event of arrest does not differ. The cover story should be simple and clear, correspond to the meeting place, the official and social status of those meeting, their personal data (age, inclinations, etc.), explain the casual or long-term acquaintance of the intelligence officer and the agent and the circumstances under which it took place. If the intelligence officer and the agent know each other through official work, the meeting can be conducted under an effective cover story for this acquaintance.

If the meeting involves the transfer of intelligence materials, the cover story must provide an explanation for the agent's availability, for example, the intention to work at home on these materials or the desire to consult with a familiar specialist, etc. The explanation must be based on the agent's actual capabilities in this regard. As a precaution, the transfer of intelligence materials to the intelligence officer should be carried out at the very last moment, before the intelligence officer separates from the agent.

At meetings that require longer-term joint presence of the intelligence officer and the agent, no cutting materials, large sums of money or operational equipment should be transferred, that is, nothing that could compromise those meeting as being intelligence personnel. The agent must be strictly warned about this. If, due to some special circumstances, the agent nevertheless brings intelligence materials to the planned long meeting, the conversation should not be held, but the material should be accepted and quickly disperse. If the location is not suitable for receiving material from the agent, it is necessary Immediately reschedule the meeting to another location and take intelligence documents from the agent.

The intelligence officer should not have with him (equally in the apartment) any records, notes, maps, documents or other materials which could indicate that he is engaged in intelligence activities (spare books with encrypted meeting conditions, telephone numbers, shingles, a list of issues to be discussed with the agent, etc.). All materials related to intelligence work must be stored in the appropriate station location.

After receiving a written report or intelligence documents from an agent, the intelligence officer must hide them in his possession, having first carefully thought out how to get rid of them if there is a danger of arrest on the way home. A small report printed on thin tissue paper or microphotographic film can be relatively easily hidden or destroyed upon arrest. If the material is in such volume that in the event of a personal search of the intelligence officer it will still be discovered, it is not recommended to hide the material and have it all in clothing. In this case, it is advisable to pack it in a bundle along with the purchase made in advance in the store, or put it in a purchased book, so that when arrested you have the POSSIBILITY of declaring that the material was planted on the intelligence officer for provocative purposes. If circumstances allow, you should try to throw away this package.

The very fact of even the accidental detention of an intelligence officer obliges him to temporarily stop meetings with the agents in contact with him until the reasons and consequences of the detention are clarified.

In case of detention and attempts at interrogation, the intelligence officer is obliged to demand that a representative of his embassy be summoned; Without this representative he should not say anything else at all.

During the arrest and investigation, one must behave in such a way that the interests of the socialist state are not harmed, the interests of the intelligence organization are not damaged, and the involvement of the arrested person in intelligence is not revealed.

Actions to Ensure the Reliability and Efficiency of Personal Communication

The Intelligence service takes the necessary measures to ensure that communications between agents are carried out reliably, uninterruptedly and quickly.

In practice, there are cases when an intelligence officer or agent cannot go to an agreed meeting or when one of them is forced to meet urgently without waiting for the agreed meeting date.

To prevent the severance of personal communication with an agent, ensuring its uninterrupted operation and efficiency in the work of the agent, the following are practiced, depending on the purpose: meetings:

Another meeting. There will be a plan for each subsequent meeting with the agent, which is one of the most common ways of maintaining communication, giving the opportunity to the intelligence officer to personally supervise the work of the agent and his training and evaluation. Regular meetings are held as necessary between an intelligence officer (cell leader) and an agent known to each other.

The interval between regular meetings depends on a number of factors, including the position of the agent, the nature of the tasks he performs, and the intelligence and operational situation.

The elements of a regular meeting are: date (day) of the meeting, time of the meeting and place.

The agent's attendance at the meeting is mandatory. However, in practice there are often cases when an agent violates this provision for various reasons: illness, going on a business trip, being busy at work, discovering that he is being followed, etc. Sometimes an agent does not show up for a meeting as a result of an unclear communication agreement or goes to a meeting, but at a different location. In this regard, you should arrange a backup meeting with the agent.

Spare meeting. This meeting with the agent is determined by the intelligence officer in case the next meeting is disrupted.

The main purpose is to ensure uninterrupted communication with the agent. A backup meeting, depending on the agreement, can be held a few hours later, the next day, or a few days after the next meeting. The intelligence officer agrees on a backup meeting with the agent when he schedules his next meeting.

Sometimes they arrange not one, but several backup meetings. A place, time and day for a backup meeting are set. In some cases, to make it easier for the agent to remember the terms of communication, the place and time of the backup meeting remain the same as for the next one, but this depends on specific circumstances. For example, you cannot go to the same movie at the same time two days in a row if the meeting place is fixed at some cinema, and the time of the reserve meeting is the next day. It also happens that the reason for the breakdown of another meeting may be the current situation directly at the meeting place.

An agent or intelligence officer may, for example, notice something suspicious at the meeting place. Then, naturally, a backup meeting between the intelligence officer and the agent can only take place if it is arranged at another place.

Consequently, the conditions for a reserve meeting should be developed as carefully and thoughtfully as the conditions for the next meeting.

In the process of intelligence work, there may be a need to urgently hold an extraordinary meeting between the intelligence officer and the agent. In such cases, a so-called emergency meeting is used.

Emergency meeting. Such a meeting between an intelligence officer (or cell leader) and an agent or illegal immigrant is carried out on the issue of one of the parties with the help of a pre-developed and conditioned, unmistakable and strong signal to resolve suddenly urgent problems in intelligence activities.

For example, an intelligence officer may need to transfer an urgent task to the agent, warn him about the danger of compromise, the need to take appropriate measures, etc. the agent, in turn, may have urgent questions that need to be quickly reported to the intelligence officer, for example, he has received intelligence materials that need to be urgently transferred, wants to warn intelligence about the danger of an upcoming enemy operation, etc. In such a case, it is necessary to determine the emergency place and time of the meeting and develop a method of calling.

The place and time for an emergency meeting may not be related to either the regular or the alternate meeting, but the places and time of both meetings can be used for this purpose. It is important that the agent is able to remember these conditions and not confuse where and when to go to meet during an emergency call.

A call to such a meeting is carried out in various ways, depending on the conditions in which the agent and intelligence officer live and work, and on their ingenuity. This could be an object-graphic signal placed somewhere or a signal given on the radio, a predetermined telephone call, sending letters, postcards, telegrams by mail, placing an advertisement in a newspaper, etc. It is important that all this is done strictly in compliance with secrecy. It should not be allowed for a reconnaissance agent to use the same place to set up a signal for a long time, since this could lead to the enemy's counterintelligence deciphering the communication between the intelligence officer and the agent.

When the intelligence-operational situation in the country does not allow frequent personal meetings; so-called instant meetings are used, which significantly reduce the danger of deciphering the contact between an intelligence officer and an agent.

An instant meeting is a secret meeting in which those meeting, without giving the appearance that they know each other, enter into a conversation, but using suitable moments and designated places, unnoticed by others, they transfer secret materials. Such meetings are used only for the rapid transmission or reception of intelligence material. Transferring classified materials in an instant meeting is a complex undertaking. Of primary importance here is the selection of a place that should be convenient for the operation and guarantee its safety. When choosing a location, you should strive to ensure that at the time of the operation it cannot be in the field of view of the external surveillance operatives. In this case, the synchronicity of the actions of those meeting plays a big role.

The transferred material must have reliable camouflage and packaging convenient for sending and receiving. These meetings must be held taking into account the characteristics of the agent tour and operational situation in the reconnaissance country and specifically on the spot, as well as taking into account the social and official status of those meeting. During an instant meeting, it is very important to maintain discreet behavior of those meeting so as not to attract the attention of random witnesses.

The elements of an instant meeting, like other meetings, are the date, time and place of the meeting. To avoid disruptions to instant transmissions, it is recommended, where necessary and appropriate, to have a backup location and time for instant transmissions. Time gap between the main and backup options should be small. When intelligence materials are exchanged often, for example, an agent regularly transmits information on film, it is advisable to negotiate with the agent on the terms of urgent call for a meeting.

A visual meeting is a meeting in which those meeting do not come into direct contact, but are explained in clear terms with the help of pre-arranged signals.

Some examples of such pre-arranged signals are as follows:

- *"I need to hand over the materials"*, - the newspaper is in my right hand,

- *"Things are going well"* - I am wearing a red tie,

- *"I received the telegram"* - I stop to adjust my shoelaces.

- *"I understood the task"* - I rub my hands together three times etc.

Signals should be chosen that would not attract the attention of enemy surveillance operatives.

The elements of a visual meeting are the meeting date, time and place. When properly organized, this meeting is one of the safest. The farther people are from each other, the more difficult it is for counterintelligence to record the meeting.

Visual meetings can be triggered for various reasons. Such meetings are mostly held by agents who have temporarily stopped active intelligence activities due to the threat of compromise or are not used in their work for other reasons. This type of meeting often serves as an auxiliary means to facilitate work on other communication channels.

Due to visual meetings, it is possible, for example, to reduce the number of other meetings if, at a visual meeting held some time before the scheduled next meeting, a signal is provided about the inappropriateness of its holding or its postponement to another time.

Often a visual meeting is a component of any other meeting. So, before carrying out an instant transfer, those meeting first establish visual contact, after which they follow to the place of transfer. At the next meeting, it is also practiced to pre-establish visual contact, which makes it possible to cancel the meeting in case of danger. Visual meetings are also held when it is necessary to determine an agent's status.

Permanent Conditions of Communication

The meetings discussed above as ways to maintain contact with an agent ensure uninterrupted communication, but they do not guarantee us against the possibility of an agent or an illegal alien losing contact. There can be different cases: the intelligence and operational situation suddenly became complicated, the agent messed up something in the communication conditions, the intelligence officer urgently left the country, etc. The agent remains without communication. Due to his inexperience or ignorance, he may take steps to restore this connection.

Often such steps lead to unnecessary complications in working with the agent, and sometimes to the compromise of the agent.

To avoid this, it is necessary to agree with each agent on permanent communication conditions that ensure the possibility of restoring communication lost for any reason. For this purpose, it is agreed with the agent that in the event of a break in communication with the intelligence officer for a long time, he begins to go to a meeting under constant communication conditions. Permanent communication conditions usually include the following elements:

Date. For example, the first Wednesday of every month or every odd-numbered months, the first Friday of every third month, depending on intelligence needs and the nature of the agent. Some number may be taken instead of the day of the week.

Place. When choosing a meeting place, you should proceed from the idea that it must guarantee the possibility of identifying the agent in case another intelligence officer contacts him, be reliable in terms of security, give the agent the opportunity to create a cover story his appearance in this place in case he has to contact an intelligence officer and meet several times in a row.

This could be a cinema, a restaurant or cafe, a metro station, a beer bar, a place somewhere on the agent's regular route, etc. in the country where the agent lives and operates, and perhaps in some other country convenient for the agent.

Time. It should be good and easy to remember (given according to local time or Greenwich Mean Time). If the meeting is arranged in a closed premises, then the time is chosen during which this premises is open to visitors. It happens that a cafe, restaurant, bar or cinema turns out to be closed on the day of the meeting for various reasons (renovation, sanitary day closed altogether, etc.), so it is necessary to stipulate that the meeting will take place on the street at the entrance to the premises.

Identifying Marks and Signs

These signs are mandatory for the agent, since, as mentioned above, an intelligence officer who does not know the agent may come to the meeting. The development of identification features requires serious attention and consideration of a number of factors including the operational situation in the country.

If, for example, in Germany the bulk of men are at work on the day and they tend to walk carrying briefcases, then the agent's briefcase of a certain color can be an identifying feature in combination with any other sign and it will not look out of place. A newspaper or magazine in his hand will look unnatural, since Germans usually carry them in a briefcase or in the inside pocket of a coat or jacket. And in the USA, a magazine or newspaper folded in half in the hand of an agent will look completely natural and can be taken as identification marks, while specifying the name of the newspaper or magazine.

Identification features can include not only objects in the agent's hands, but also features of clothing. They should be taken taking into account local habits and customs and correspond to the time of year.

Password and review is a phrase (or several phrases) determined by intelligence, any thing or item of personal use, the message or presentation of which to another person associated with intelligence indicates that the person, the password, is actually authorized by intelligence to call the planned event. The password always provides a conditional revocation.

Password and revocation as elements of permanent conditions are needed for purposes of secrecy and security. They are verbal and material connections. The verbal password and feedback are new, and the physical ones are additional, reinforcing, and provide a full guarantee that the necessary contact has been established. The main requirements for passwords are that they be simple, so that they can be recalled, otherwise it will be difficult to remember, that they correspond to the situation at the meeting place and do not arouse suspicion.

When composing conditional phrases, situations that often occur in life are used. For example, the intelligence officer might ask if they have met somewhere before. At the same time, as the password keyword, he names a specific place (city, resort, country, company, family of an imaginary friend, educational institution, etc.). The agent's review consists of affirming or denying the fact of their acquaintance using such a keyword of the review, that is, they could have met somewhere other than that (the agent names the place, which is his keyword). Often, when developing a password, we take a common street address from one passer-by to another with a request to explain the way to some place in the city and the answer to this appeal, containing a description of a certain route that differs from the usual one, that is, one that can be given to any passerby.

To restore an unexpectedly broken connection with an agent, the intelligence officer must have the most comprehensive information about him, know his exact home address, place of work, route from home to work, home telephone number, home conditions, roommates, if any, the agent's personal car, its number, that is, everything, which may help you find the agent.

Counter-surveillance

Counter-surveillance is aimed at identifying enemy surveillance of an operative or agent who is on his way to carry out an intelligence mission. It is often practiced to ensure the security of personal meetings between intelligence officers and agents. Counter-surveillance is carried out according to a pre-developed plan and route, usually carried out by several intelligence officers.

Monitoring an intelligence officer's personal contacts, especially when he works from the position of legal cover, is an effective means of identifying his intelligence activities, and therefore enemy counterintelligence pays special attention to the external surveillance service as a means of combating foreign intelligence services. The external surveillance service is constantly improving its techniques and methods of surveillance and is equipped with modern technology.

An intelligence officer must carefully prepare for each of his personal new meeting with the agent. Checking to identify the enemy's external surveillance of both the intelligence officer himself and the agent is the main and fundamental condition in ensuring the safety of personal meetings. In this difficult fight against the enemy, the intelligence officer who has good operational training, great endurance, and resourcefulness and operational invention wins.

The intelligence officer must skillfully detect surveillance and not show the employees of the external surveillance service his familiarity with the techniques of checking and evading surveillance in order not to be compromised. On the route he must behave calmly and naturally at all times. An intelligence officer has no right to come into contact with agents without being sure that he is not under external surveillance.

The complexity of the administrative-police regime in some intelligence countries often does not allow intelligence officers, especially those who do not have sufficient experience in working with agents, to independently detect external surveillance. Therefore, in "legal" Stations, in order to better guarantee the safety of meetings, counter-surveillance of the intelligence officer by other station employees is practiced in such cases.

It is more difficult to organize counter-surveillance of an agent. This is due to a number of reasons. Counter-surveillance requires precision in the agent's behavior on the route, which is associated with detailed instructions, which is not always possible to carry out at a meeting. The inspection route should be developed in advance and agreed upon at a previous meeting. However, the intervals between meetings are sometimes so long that the agent may forget some details of his route, and sometimes unforeseen circumstances arise that cannot be taken into account in advance.

Revealing even the appearance of an agent to other intelligence officers who are involved only in conducting one-time counter-surveillance is often inappropriate, and conducting surveillance only with the help of one intelligence officer who directly meets with the agent is dangerous, since this can lead to counter-productive results.

And yet, despite these difficulties, the intelligence officer, especially in difficult conditions, is forced to resort to organizing counter-surveillance of his agents in order to protect his work.

Agents must first of all be taught techniques for detecting external surveillance and avoiding it, achieving perfection in mastering these techniques.

The ability of the agent himself to detect surveillance and, if necessary, to escape surveillance is the main guarantee of his security. When intelligence believes in the sincerity of an agent and is not confident in his ability to detect himself under surveillance, it organizes counter-surveillance of him using the security personnel from the Station.

PART III

NON-PERSONAL COMMUNICATION

General Principles of Non-Personal Communication

All methods of communication that allow the transfer of operational and intelligence materials and messages without the intelligence officer and agent coming into personal contact are called impersonal communication in intelligence.

Non-human methods of communication include radio communication, communication through hiding places (dead letter boxes), by mail using secret writing and microphotography, the use of the press and other possibilities for this purpose.

The main advantage of these means of communication is that, when used correctly, they are less vulnerable to enemy counterintelligence agencies. This advantage is of particular importance at the present time, during a period of deteriorating conditions for the work of intelligence services of the countries of the socialist commonwealth in the most important capitalist countries, as a result of the intensification of the activities of police and counterintelligence agencies directed against socialist countries.

Naturally, impersonal methods of communication by themselves do not provide absolute secrecy and security of communication. The use of these methods to communicate with agents requires the same careful preparation and compliance with the requirements of communication as when making personal communications. Let's take a closer look at impersonal methods of communication.

Short-Range Radio Communication and Signaling

Short-range radio communications and signaling have recently become more actively used in intelligence work. With the power of radio equipment and other special short-range equipment, it is possible to transmit and receive intelligence information and other messages, send danger signals, call for a meeting, etc. at distances from several tens of meters to three to five kilometers.

A distinctive feature of short-range communication and signaling equipment is its independence from the local power grid, as well as its small size, which allows it to be secretly transferred to the area of intended use in clothing pockets, a briefcase, a bag, etc. Thanks to the low power of radio stations of this type and small ranges, their operation becomes practically invulnerable, since it is almost impossible for enemy counterintelligence to detect them, much less detect correspondents.

Using short-range equipment, it is possible to organize both one-way communication (by sending signals to a regular home radio receiver of the receiving correspondent) and two-way contact between correspondents located within the range of the equipment. Such contact can also be made while correspondents are moving.

Another means of two-way communication over short distances within the same transformer line is communication via electric lighting wires. With the help of special devices plugged into electrical outlets, correspondents can conduct negotiations while in different rooms of a large multi-story building or even in different houses located at a distance of several blocks from each other, provided that the electrical network of these houses is connected to one transformer substation.

Signaling is used as an additional means in almost all types of communication. Along with this, it can be used as a way of impersonal communication with agents. With a well-developed signaling system, it is possible, if necessary, to receive short messages from the agent and transmit instructions to him.

For example, a special infrared filter can be installed on an ordinary pocket flashlight, with the help of which it is possible to send signals to the agent at a distance of 3 to 5 kilometers even in the dark. Invisible infrared rays are received by the agent with the power of a special apparatus.

The development of modern technology makes it possible to use other means for impersonal communication with agents.

Communication Through Dead Letter Boxes

In intelligence, a dead letter box is a specially selected and sometimes adapted place for the secret storage and transfer of operational materials. Dead letter boxes can be used for impersonal communication between all intelligence units.

Through dead letter boxes, both small-sized intelligence materials (photo films, microdots, small packages) and materials with large dimensions (radio and photographic equipment, samples of new equipment, drawings, models) can be transmitted. Some dead letter boxes may contain operational equipment, materiel, and money. Therefore, the choice of location for a hiding place and its equipment is made in accordance with its intended purpose.

In order to ensure the security of communications between intelligence units when using a dead letter box, it is necessary that the dead letter box meet the requirements of secrecy.

When looking for a place for a hiding place, one should take into account the intelligence and operational situation: the administrative and police regime, the characteristics of the city or region, the life and customs of the local population.

Knowledge of the intelligence and operational situation makes it possible to show great ingenuity in finding the most suitable places for hiding places, to use the characteristics of each country, each region and city, and the way of life of the population. Depending on these features, dead letter boxes can be located in parks, squares, on the side of rivers, roads, in the foundation and in cracks of stone fences, in destroyed and intact buildings, in stadiums and sports grounds, hippodromes, in public restrooms, etc.

Hiding places can be arranged in a wide variety of places and the task of the intelligence officers is to ensure that such places in which the planted material would not be accidentally discovered by an outsider, so that it would be easy for the intelligence officer and agent to secretly plant or seize the material. For this reason, hiding places should not be installed near protected objects, restricted areas, stationary counterintelligence observation posts, or military posts. The approaches to the dead letter box must be clearly visible, and the dead letter box itself must be located in a place that would allow viewing the surrounding area and exclude the possibility of outside observation of the actions of an intelligence officer or agent while using the dead letter box.

When choosing a location for a dead letter box, you must take into account public and official position of agent and intelligence officer. The dead letter box should be set up in a place where visiting would be quite natural for both the intelligence officer and the agent. Otherwise, the presence of the intelligence officer and agent in the place of walking around the dead letter box may arouse the suspicion of others and the contents of the dead letter box may become the property of counterintelligence.

When choosing a location for a dead letter box, you must also take into account what time of year this dead letter box will be used and what the climate characteristics of the country being explored are. Dead letter box, suitable use in winter conditions, and vice versa. When using a dead letter box in winter, you need to take into account the presence of passers-by, This may lead to the disclosure of the dead letter box. Snow cover on the approaches to the dead letter box in winter can compromise the dead letter box.

In winter, you should not set up a dead letter box in a place located away from the road, where no one usually walks, since after an intelligence officer or agent visits the dead letter box, tracks will remain in the snow that can attract the attention of the police, janitors, watchmen, or simply curious passers-by. Better to select a dead letter box site during winter that sees enough use that tracks in the snow in the area won't appear suspicious.

In summer, the same site may be unsuitable for various reasons, so the intelligence officer should always select a new dead letter box from scratch. A hiding place, located, for example, in a country park, stadium or sports ground, can be successfully used in the summer. In the fall, when the rains begin, a visit by an intelligence officer or agent to these places will not be justified and may arouse suspicion.

When arranging a dead letter box, you should also take into account that the process of inserting and removing materials. This process of filling and emptying the dead letter box should be simple, convenient, not require special physical effort, not require special tools and not take a lot of time. Otherwise, the moment of insertion or removal of materials may be detected by enemy counterintelligence or third parties.

Hiding places should not be established in items that are subject to movement or can be picked up by unauthorized persons, for example, in empty boxes placed as unnecessary in the yard or in some other place.
.

The dead letter box should have good landmarks so that it can be easily found. Its description should be concise and precise. It is also good to describe the dead letter box graphically. This is especially important when communication through a dead letter box must be carried out by a messenger or courier who is not familiar with the area.

Below is a description of several hiding places that are now known to American counterintelligence, but they were used by one of the intelligence services of the socialist countries.

1) *"New York, Manhattan, Riverside and 96th street, men's room at the Playground. The far right booth, if you are facing the booths. From a sitting position inside the booth, use your left hand to attach a magnetic container inside the only pipe there to which the cabin wall is attached. The size of the container is 2x3x10 cm."*

2) *"New York, Manhattan, Symphony Cinema on Broadway and 25th Street. Upon entering the auditorium, go right, go up the stairs to the balcony. The far right seat of the last row of the balcony. When an intelligence officer or agent sits in this chair, they discover that there is no need to bend over for the container, the floor is carpeted at elbow level, the carpet is nailed to the floor, and in this place it is slightly torn off the floor so that a bookmark can be made. From the "sitting" position in the chair", place right hand under the carpet, opposite the chair. The containers are flat."*

3) *"New York, Manhattan, end of Amsterdam Avenue. Walk to the sports field; from the position "facing the East River" Off the right side of the site there is a path going down through the park. You need to go down this path with a guide to three pipes, located on the opposite side of the river, to a large dry stump, which will be on the left side of the path. The roots of this stump serve as a dead letter box. The container is placed in the roots under the stump on its left side, if you sit on the stump and look down at the river."*

4) *"New York, Central Park, a lake with a boat station. In its northern part, where a stream flows in, there is a small wooden bridge across this stream. Standing on the bridge, facing the lake, and leaning on the railing , a pin-type container is pinned under the railing strictly in the center of the rivulet, in the place where the angle is formed between the top board and the railing beam."*

After the dead letter box is constructed and adapted for use, it is necessary to test it in the field. The dead letter box is tested by placing an object in it. In this case, you need to see if the object and the hiding place are visible after emplacement. If after a few days when checking the location, the dead letter box and the item it contains are still there, intact and not tampered with, then we can assume that dead letter box is suitable for commissioning. When testing dead letter boxes, you cannot use objects that may attract attention, since if a person discovers it, he will at best he will take it for himself, and at worst he will report it to the police, and the site will be compromised from the very beginning, even before it is put into operation.

Such hiding places as a means of impersonal communication are used not only by intelligence organizations, but also in a number of other illegal activities. For instance, children often use them in their games.

In the conditions of New York, there was a case when one intelligence officer, while selecting dead letter boxes, discovered a small bag in one convenient place for a dead letter box. From a position sitting on a bench, he tried to pull it out of its hiding place, but it turned out to be heavy. He did not dare to withdraw this deposit, he went to the station and reported this to the resident station chief. An hour later, when two intelligence officers returned to the area to inspect the contents of the bag, the hiding place it was no longer there. Therefore, hollows in trees, caves in the mountains and other secret but obvious storage facilities located in plain sight by passers-by are not suitable as hiding places for intelligence purposes.

The dead letter box should not attract the attention of others nor should the container placed in it. It is always advisable to transmit information in film or microdot which are both easily hidden. There may, of course, be exceptions caused by specific circumstances.

If the material in the dead letter box must be kept for a limited time (no more than an hour) and trash items are used to disguise it, then the place for the dead letter box does not have to be selected in some hidden recess. In this case, the container, which is a discarded item suitable for the area, can be placed near a landmark. However, such places must also meet the requirements for to all hiding places.

Each dead letter box can only be used for communication by one agent. It is strictly forbidden to maintain contact with several agents through the same dead letter box, as this can lead, firstly, to agents deciphering each other, and secondly, to the compromise of the dead letter box itself. For example, if the dead letter box site is located by hostile, counterintelligence, they can place it under close surveillance and will eventually identify all the persons using it.

The dead letter box serves its purpose only when it looks natural. For example, one young illegal intelligence officer set up a hiding place in the forest, using a liter wall jar with a screw cap. He dug a jar into the ground, covered it over and placed a small stone on top. The result was an ideal hiding place - a jar under a stone. The material was placed in a jar and hermetically sealed, thus preventing it from being damaged by rain and moisture. Despite the simplicity, such a hiding place turned out to be perfect for operational use, since the jar remained well hidden under a stone the whole time. However, if the operatives of the enemy surveillance service had even accidentally recorded the action of the intelligence officer (the latter could most likely have been under surveillance if he was acting with legal positions), then they would have discovered the jar found under a stone,

A natural dead letter box is good because after removing materials from it, there are no traces left in it, and if enemy counterintelligence begins to control such a dead letter box, even for preventive purposes, it will not give up anything, not even fingerprints. The enemy counterintelligence organization does not know when this dead letter box will be used next time, or indeed if it is a dead letter box at all, but they cannot risk it, so they will need to waste resources and manpower keeping the site under observation.

It is not recommended to make hiding containers in public places from objects found there (door handles, handles of toilet tanks, ozonizers, etc.) and leave them there. Such secret containers should be in place only when there is an investment in them, that is, the time required for the secret operation. When the dead letter box is not in use, it should contain normal branded items that will not cause any suspicions.

Dead letter boxes can be categorized into closed and open.

Closed dead letter boxes are those that are located in a room where entry is restricted to certain times of the day or night. Open dead letter boxes are dead letter boxes located in the open air in parks, forests, on the street, etc. According to their purpose, dead letter boxes can also be divided into small-sized and voluminous, simple and magnetic.

The highest level of security is achieved by maintaining communications through a dead letter box when it is used only once. This is possible in conditions where the exchange of information through a dead letter box is episodic (one-time) in nature. It is impossible to achieve one-time use of dead letter boxes if the agent provides ongoing materials over a period of time. In that case, a certain system is required that would provide secret, reliable and operational communication through hiding places. This system can be varied depending on each case. The stencil template is dangerous here too.

There is, however, a principle of working through dead letter boxes, which is known, perhaps, to all intelligence services: this is the consolidation of several dead letter boxes into one system. A certain number of dead letter boxes are selected taking into account the needs of reconnaissance; the dead letter boxes are numbered, for example, from the first to the tenth or from the first to the fifteenth, and in this quantity they are immediately put into operation.

To call to these dead letter box hiding places, a system of signals was developed that makes it possible to call an agent or intelligence officer to a hiding place at any time using certain signals. Signals can be a variety of visual signals (subject or graphic), on the radio or by telephone, a signal can be an advertisement in print, etc. But all these signals must provide for at least three operations:

1. *the dead letter box has been placed;*

2. *the contents of the dead letter box have been safely retrieved;* and

3. *the dead letter box has been compromised or abandoned.*

The need for the first signal is obvious, the second signal must be given to confirm that the contents of the dead letter box have been received safely and taken away, the third signal about closing the dead letter box is given in cases where the item in the dead letter box has not been located, and when some suspicious issues have arisen related to the safety of work through this dead letter box, when they simply decided to abandon this dead letter box for some other reason reasons or when the dead letter box is lost.

The one who gave the signal to abandon the dead letter box is obliged to inform the other party as soon as possible about the reasons for this signal, select a new dead letter box and continue working if there is no other agreement.

When using graphic signals, it is necessary to select two places for their placement. One location is somewhere on the daily route of the agent, which is used by the intelligence officer to call the agent, the second on the daily route of the intelligence officer or other trusted person, which is used by the agent to call the intelligence officer to the dead letter box. After reading, the signals are destroyed by the party that placed them. Otherwise, it is easier for counterintelligence to identify the parties involved in organizing communications.

Putting materials into a dead letter box and their removal, setting various signals related to work through a dead letter box are the same responsible operations as personal meetings with agents, and therefore require compliance with all the rules of secrecy is necessary

The absence of external surveillance along the route to the dead letter box does not yet provide a guarantee of security, since counterintelligence, if it discovers a dead letter box, can organize in advance a hidden external surveillance device in the area of its location. Therefore, it is necessary to especially carefully check the area of the hiding place, study the surrounding area and pay special attention to those premises or areas that counterintelligence could use for a hidden observation post.

It should also be borne in mind that some counterintelligence agencies use sniffer dogs to detect hiding places. This obliges the intelligence officer and agent to take measures to confuse their tracks when going to the dead letter box and when returning from the dead letter box, using various types of transport, visiting hungry places, etc. In order to prevent the danger of counterintelligence or random persons reading information transmitted through an agent's dead letter box,

When reporting or intelligence instructions, maximum use must be made of operational techniques, secret writing, microphotography, as well as numbers and codes.

For these purposes, special packaging is sometimes used, when opened by an unauthorized person who does not know the secret of the packaging, the material contained in it is automatically destroyed or damaged.

Great care should be taken to disguise materials transferred through the dead letter box. Methods of camouflage can be very diverse: an old box of matches or cigarettes, a tin can, a piece of crumpled newspaper, a piece of pencil, a cigarette butt, etc. If the dead letter box is located in the soil, it is convenient to use a piece of metal tube, pointed at one end so that it can be easily pressed into the ground.

It is not recommended (except in special cases) to transmit through the dead letter box materials relating to the organizational work of intelligence, installation data and addresses of agents, safe houses, instructions on appearances, meetings.

In cases of extreme necessity, this kind of information and even personal documents for illegal immigrants can be transmitted through a dead letter box, but the dead letter box operation must be carried out in such a way that the contents in the dead letter box are not lost from view. To do this, there must be an exact agreement on the time of placing and removing material. The person who deposited the material then monitors the dead letter box until the seizure is recorded.

A visit to the location of the dead letter box by both the

intelligence officer and the agent must be justified by some plausible pretext. In case of police arrest or questioning of acquaintances who meet by chance, a cover story must be developed to explain the presence of an intelligence officer or agent in the area where the dead letter box is located.

When communicating through a dead letter box, the persons carrying out it have no the need to know each other. Therefore, if intelligence, for one reason or another, does not want to reveal to the contact agent the agent to whom he is directed (or, conversely, to the agent - the contact person), communication between them can be organized through a hiding place. When using a dead letter box, you can, if necessary, successfully replace one intelligence officer with another or even someone involved in cooperation with intelligence, because to remove and invest materials in a dead letter box, knowledge of neither the agent, nor the nature of the work carried out with him, nor the corresponding foreign language.

The use of impersonal communication means, however, creates difficulties in managing and training agents. The lack of personal contact with an agent when using a dead letter box deprives the intelligence officer of the opportunity to constantly observe the agent, monitor his moods, promptly exert moral and political influence on him, systematically improve his intelligence skills, more fully identify his agent capabilities, then is complicates the operational management of the agent's work, its study and verification.

This circumstance must be taken into account when working with newly recruited agents, who need to be especially carefully studied and checked and who need guidance and education. It is necessary to conduct personal meetings with such agents more often.

The disadvantages of transmitting intelligence materials through dead letter boxes also include the fact that intelligence loses these materials out of their hands for some time. If a dead letter box is discovered by enemy counterintelligence, the enemy can be tracked down and arrested.

With the skillful and subtle work of enemy counterintelligence, it is a fact the arrest of an agent may not be detected for a long time, disinformation materials can harm the intelligence of a socialist state. There have been cases where enemy counterintelligence have compromised, and captured an agent and coerced him to work for them. The agent and his dead letter boxes were then used to transmit disinformation.

This type of situation occurs when the enemy forces discover a dead letter box is in use, then organize an ambush at the discovered hiding place in order to capture the intelligence officer.

The enemy can also then simply use the circumstance of the captured agent and any materials recovered from him in propaganda against the socialist state.

In recent years, in the practice of Stations in countries with a difficult intelligence-operational situation for the intelligence services of socialist countries, one-time-use dead letter boxes are increasingly being used, which make it much more difficult for enemy counterintelligence to open and intercept this communication channel.

One-time hiding places, as a rule, are selected only by the party emplacing the dead letter box. The person making the retrieval usually does not check these dead letter boxes for the purpose of tradecraft and only goes out to retrieve materials, using the description of the dead letter box. Much attention is paid to embedding the transferred materials in containers disguised as various waste items or in magnetic containers.

Emplacements and retrievals are strictly specified in order to minimize the time the materials remain in the dead letter box and thus reduce the risk of its accidental discovery by unauthorized persons. Insertion and retrieval signals are sent via radio and other means.

When working with one-time dead letter boxes, you must have several hiding places, a clear system for their use and signaling to them. When working regularly through a system of one-time dead letter boxes, a schedule for their use is developed, that is, a certain time for inserting materials and their removal is determined in advance.

Exploiting the Press for Agent Communications

One of the auxiliary means of impersonal communication with agents is exploitation of the press in the country in which the intelligence service is operating.

The essence of using this means of communication is that in newspapers, magazines or other periodicals pre- arranged between the agent and the intelligence officer, the agent or the intelligence officer places coded messages, for example, under the guise of an article on a certain topic or an announcement.

Placing the article naturally allows for a broader message to be conveyed and makes coding easier. However, it is difficult to use this means if the intelligence service does not have its own agent on the editorial board of the newspaper or magazine. It is not always possible to obtain editorial consent to publish an article. The article itself may undergo certain editorial changes and abbreviations, which will disrupt the content and in some cases may make it impossible to decipher the transmitted message.

The publication of advertisements for the transmission of coded messages is much more accessible, since bourgeois newspapers willingly publish a wide variety of advertisements from both private individuals and various organizations, firms, enterprises.

The use of this means of communication requires knowledge of the nature of newspapers published in a given country in order to identify the most widespread ones and those that print the largest number of advertisements. Newspapers such as, for example, the Washington Post, the New York Times, and the Times publish hundreds of advertisements every day on issues of labor supply and demand, apartment rentals, and the purchase and sale of cars and other property.

Different countries, and even cities, have their own

procedures and rules for publishing advertisements, which you need to know before sending an advertisement to the newspaper. Newspaper advertisements can be submitted to the editorial office in person or by mail with the condition that they be printed the next day, in a few days, with a request to print these advertisements within a week, daily or every other day.

As a rule, the editors do not require any documents identifying the person submitting the advertisement. In published advertisements, you do not have to indicate the address of the person who placed the advertisement; if necessary, you can refer to the receipt number. The certificate issued upon acceptance of the advertisement for publication, will usually simply give the address of the editorial office rather than the address of the person or company buying the advertising..

Some newspapers and magazines require that the advertiser indicate his address. In such cases, you should place the ad through a third party or state that the person placing the ad is passing through the city and does not have a permanent address in the city.

When using this means of communication, it must be borne in mind that hostile counterintelligence agencies are usually interested in newspaper advertisements in order to identify suspicious messages. This obliges intelligence officers to systematically and carefully study the content and form of newspaper advertisements so that the nature of the message placed by intelligence does not differ in any way from ordinary publications. It is not recommended to report the loss or disappearance of any items in such announcements, as this may attract the attention of the criminal police. This format is only suitable for a one-time message. Ads should not convey anything particularly original that might be of interest to counterintelligence agencies or police officers.

At the same time, the announcement must be made in such a way as to at once so that it can be easily identified by the person who is to receive the message.

For example, in one of the capital newspapers of a Balkan

country, the following announcement was placed in the chronicle section:

> "The director of an English trading house has been in our capital for several days... Tomorrow he leaves for Istanbul, where he will stay for 5 days. Those interested in the range of goods this house sells can find it at the following address..."

By placing such an advertisement in the newspaper, the British intelligence courier made himself known to the agent to whom he came to communicate.

The publication of various announcements can be used by intelligence to transmit signals about the arrival of a courier, a call to an urgent meeting, a warning about danger, and to transfer brief items of information of intelligence value.

The use of various publications in the periodical press as means of communication is constrained by certain limits: in this way only short messages can be transmitted, and not often, since the regular publication of the same type of announcement may arouse the suspicion of enemy counterintelligence agencies.

The use of this method of communication is advisable mainly for communication with agents operating at the station location, and not only for communication with agents within the intelligence country, but also for communication with agents located in other countries, since more and more smaller periodicals are widely distributed abroad.

Thus, the use of the press to communicate with agents is only considered an auxiliary means that can be effectively used in combination with other means of communication.

It should be pointed out that the distribution of press and other printed publications can be used as a means of sending intelligence messages within one city and to other cities and countries. In this case, printed publications are used as material on which intelligence officers and agents write their messages in secret or transmit them by means of certain marks, for example, by pricking letters in the relevant articles. Press materials can be sent by mail or placed directly in recipients' personal mailboxes.

PART IV

COMMUNICATION USING INTERMEDIARIES

General Principles of Communication Using Intermediaries

An intermediary connection is usually understood as a connection carried out between an intelligence officer and an agent with the help of third parties, that is, intermediaries, in which reliable and proven agents are used.

Constant personal communication between an intelligence officer and an agent, which ensures the greatest efficiency, is not always possible, and in some cases may be undesirable. In view of this, intelligence sometimes organizes communication between the intelligence officer and the agent through an intermediary agent. Introducing an intermediary agent into the communication line of a third party is undesirable from the point of view of tradecraft, since this introduces an extra link into the connection and the intelligence materials of the agent or the instructions of the intelligence officer are transmitted not directly from one to the other, but through a third party. However, skillfully organized intermediary communication under certain conditions, which will be discussed below, can be effective.

The reasons that force an intermediary link between the intelligence officer and the agent can be very diverse. Intelligence agents working under the guise of the official representative of the socialist state are, as a rule, subject to intensive external surveillance for a long time by local counterintelligence. Under such conditions, personal meetings with agents are difficult, and with some agents they are completely excluded. In this case, in addition to using impersonal methods of communication, it is possible to find an intermediary agent, for whom, as a local citizen, it is easier to maintain contact with the main agent, and for the intelligence officer, in turn, it is easier to organize communication with an intermediary agent who has a convenient cover.

Sometimes it is advisable to communicate with an agent through a third party if he occupies a prominent official or public position, which is very different from the position of an intelligence officer of a "legal" station, and cannot, due to specific circumstances, make an impersonal connection.

In a number of cases, it is possible to organize communication with an agent working at a particularly secret facility, which is under enhanced counterintelligence surveillance, only through an intermediary agent. There are cases when it is necessary to organize communication between the station and agents located on the periphery of the country where intelligence is being collected. The agents do not have the opportunity to travel, and the agents cannot send intelligence officers to the cities where they are located.

Intermediary agents are used to provide a station with illegal Stations and safe houses, to organize communications between illegal intelligence officers and agents. They are also use when an illegal intelligence officer or enemy defector, due to the nature of his employment, is not in a position to be able to communicate directly, either personally or non-personally with an agent.

An intelligence officer selects agent-intermediaries from existing agents or recruits them from among persons who have convenient professions that allow them to come into contact under plausible pretexts. For example, it is not particularly difficult to objectively keep counterintelligence suspects from the contact of agents and intelligence officers of socialist states with intermediary agents working as taxi drivers, librarians, gas station attendants, hairdressers, theater box office attendants, staff in pharmacies, shops and various workshops.

An intermediary agent performs important functions in intelligence. He is entrusted with an exclusively secret and complex area of work - communications. Therefore, the intermediary agent must be, above all, honest, reliable and loyal. We should always strive to ensure that ideological and political motives underlie the cooperation of an intermediary agent working for our intelligence apparatus.

An indispensable condition for intermediary agents is that they should not be compromised before local authorities by their progressive views, activities or connections with progressive circles and should not be under the suspicion of the police, security or counterintelligence authorities for any other reasons. Each station and the Center must continually engage in the issue of selection and training of intermediary agents, and thoroughly investigate the persons intended for recruitment for these purposes.

The selection of such agents should be targeted, taking into account the specific tasks of organizing communications with a particular agent. For these purposes, you can also use agents that have been tested in practical work, and have proven themselves in their knowledge, commitment, honesty and loyalty, but have been transferred to a reserve or retired status for example, due to age or the loss of agent capabilities. Some of these agents may have convenient cover, while others the necessary cover can be created.

The categories of intermediary agents include:

Liaison officer - an agent who carries out intelligence tasks without maintaining secret communications between agents and intelligence officers mainly on the territory of the target country or directly with the Center;

Agent courier - agent used by our organization to maintain secret communications with intelligence agents or individual agents located in third countries or in the same country, but at a considerable distance from the station.

In some cases, the station communications officer is authorized to, within certain limits, discuss operational issues with the agent courier orally at meetings, listen to the agent courier's opinion on a number of important issues, make decisions, which he is obliged to coordinate with the intelligence officer before implementing them. As an agent the courier does not have such powers, but mush have any concerns or problems cleared by the station communications officer before an operation.

The functions of the agent courier are to receive secret mail in a closed form (coded, encrypted and in containers) from the agent or intelligence officer and deliver this mail to another intelligence officer or agent

The courier knows nothing about the agent. In most cases, he maintains a non-impersonal connection with both sides, mainly through dead letter boxes, but a transfer point can also be used here. The agent courier, as a rule, must know about the data that it needs to work. There are cases when the liaison officers know the real position of the agent (surname, place of work, addresses, etc.). But the agents should not know the real position of the liaison if this liaison is not a personal friend of the agent, recommended to us by the agent himself.

The owner of a secret transfer point is also an agent. He owns an address or other location through which secret intelligence materials are transmitted from the source agent to the intelligence officer and from the intelligence officer to the agent;

The owner of a clandestine postal address is an agent who has a convenient cover for conducting mail correspondence, to whose address secret mail is sent for transfer to intelligence;

The owner of a clandestine telephone service is an agent whose phone in an apartment or enterprise is used by intelligence to communicate with another agent;

The owner of a clandestine safe house is an agent whose house, apartment or other building is used for clandestine communication or other intelligence purposes.

Liaison Officers

When organizing communication between an intelligence officer and an agent through a liaison officer, it is first of all necessary that the liaison officer skillfully disguise the performance of intelligence functions with his official affairs. It is advisable that the liaison officer's cover allows him to come into contact with such persons who could easily be approached for legal and justifiable reasons by both the intelligence officer and the agent between whom the liaison officer communicates.

Finding a form of communication between two local citizens (a contact and an agent) is not so difficult, even if they actually do not have natural pretexts for personal contact. Such conditions can be created, and the enemy's counterintelligence will have no reason to suspect the existence of an intelligence connection between them. It is more difficult, of course, to organize contact between an agent and an intelligence officer, and in such a way that this contact, if noticed by counterintelligence, would not arouse the enemy's suspicions.

For a liaison officer who communicates between an intelligence officer and an agent living in the same city, successful covers can be working as a taxi driver, a home delivery person, a store messenger, a laundry messenger, a postman, a cashier, an attendant of the theater box office, or the cloakroom of the theater, cafe attendant, restaurant waiter, etc. since in this case the contact of the intelligence officer with the contact person, as well as the contact of the agent with the connection person, is well covered up by completely natural circumstances..

A liaison officer communicating with an agent, living in another city of the intelligence country, must have business reasons for traveling around the country, including to the meeting place with the agent. This frees the liaison officer from the need to constantly explain his absences from family work and explain how and where he gets money for trips.

Frequent travel of a liaison officer around the country, not arising from the nature of his occupation, may attract the attention of local counterintelligence.

A very convenient cover for such a liaison can be the profession of a traveling salesman, an employee of railway, river, sea and air transport, a driver or conductor of a long-distance bus, or a newspaper correspondent.

The liaison officer may reside anywhere in the country, but

they must have a legal reason to visit the places where he comes into contact with the agent or intelligence officer.

Here is an example of the liaison officer's work. We shall for the sake of this example, assume that the liaison offer is named Nick.

To make contact with both an intelligence officer and an agent, liaison officer Nick uses various methods of communication. He can personally meet with both, use a dead letter box, a transfer point, a safe telephone and other methods.

The communication methods used by the liaison officer must be developed differently, imaginatively and based on operational tasks and specific conditions. When giving a task related to a communications officer's trip to a meeting, the intelligence officer must carefully instruct him on the following basic issues:

a) Familiarize the communications officer with the intelligence and operational situation at the point where he is sent to communicate with the agent;

b) Instruct where and how he should meet with the agent, what instructions he should give him, how and in what form he should receive the materials, how he should store them before delivery intelligence officer;

c) If the liaison officer does not know the agent by sight, the intelligence officer must inform him, in detail, of the agent's appearance;

d) If personal meetings between the intelligence officer and the agent are not provided for rush, the intelligence officer must indicate where and how the signalman should receive intelligence materials and how to transfer them the intelligence agent's instructions (for example, through a dead letter box);

e)　　If necessary, the intelligence officer must, together with the liaison officer, develop an appropriate cover story for the meeting with the agent and ensure that the liaison officer understands this cover story.

f)　　Warn the liaison officer about the strict observance of the confidentiality of the purposes of the trip and warn him against such mistakes as casual acquaintances, using hotels of a dubious reputation, infatuation with women, addiction to wine, etc.;

g)　　Teach the liaison officer how to check for and detect external surveillance.

Depending on the nature of the task, the conditions for its implementation and the personal qualities of the liaison officer, it may be advisable to remind the liaison officer how to behave in the event of an accidental or deliberate detention, search or arrest and how to get rid of materials if there is a threat of their seizure by counterintelligence; how to behave if the materials do end up in the hands of counterintelligence, and how he should behave under interrogation.

For the purpose of secrecy, the intelligence officer and agent must transmit intelligence information to the liaison officer in encrypted or coded form, and also use secret writing and use micro-points.

The materials must be anonymised so that it would be impossible to identify the source of information from them if they ended up in the hands of enemy counterintelligence.

The materials must also be disguised in order to hide the presence of them in the liaison agent's possession from others. The camouflage method is only reliable when it corresponds to the situation in which the liaison officer performs the task, when the materials are given a natural appearance, common in the given conditions. Compliance with this condition for masking materials applies to both the liaison officer, the courier agent, and the transfer point. Wherever possible, it is advisable to use the transfer of materials in microdots, which can be easily disguised in a particular item transmitted through a messenger. You don't have to tell the liaison officer about this.

To check the reliability of the communication of the liaison officer, the intelligence officer needs to find the opportunity to occasionally meet with the agent himself in order to make sure that his instructions and money are correctly transferred to the agent, and that there is a correct relationship between the liaison officer and the agent.

The intelligence officer must always be able to establish contact with an agent rather than through a liaison officer than the contact person.

Couriers

The intelligence organization uses couriers to transport large volumes of intelligence materials, documents, money and special equipment. Such official positions are recruited as agent couriers, whose profession or related connections provide them with the legal opportunity to make regular trips to the country of deployment of the Station, intelligence group or agent with whom the Center establishes contact.

A very convenient cover for a courier can be trade, work as a traveling salesman, service in a commercial or passenger fleet, on the crews of international aircraft, on international trains, work as a diplomatic courier in a target country, etc.

For a one-time trip, the courier can use any of the listed covers or special measures can be developed for him to organize this trip: the agent courier can be included in the delegation traveling to the country of interest to intelligence, he can go there on commercial matters, for family or another question, having, of course, some valid reason for this.

Circumstances may also arise when intelligence is forced

to use a courier that actually does not have pretexts for obtaining a visa to enter the country where he is going. In this case, the courier is provided with the appropriate forged documentation giving him the right to enter the country of interest.

Organizing such trips usually involves great risk. Our intelligence organization also resorts to the illegal transfer of couriers across the "green border". To do this, it is necessary to carefully study the border area and identify those sections of the border where the border security is weakest and where the conditions are most favorable for crossing.

To successfully organize communication through a courier, you need to know the intelligence and operational situation in the country where the courier is being transferred. This is necessary for the correct choice of the method of transporting the courier, drawing up a cover story and providing the courier with the necessary documents, instructing the courier on his line of conduct, as well as choosing the most appropriate way to establish contact with the intelligence officer and agent.

The courier may communicate with the intelligence officer or agent through personal meetings or the use of impersonal means of communication. Depending on what method of communication is envisaged, the station must carry out preparatory measures, for example, select hiding places, transfer points, places for appearances, and also timely develop conditions for using them.

The most appropriate form of communication between the intelligence officer and the arriving courier is a dead letter box or transfer point. The courier is informed of the location of the dead letter box or transfer point and the conditions for their use. This method of communication avoids revealing to the courier, other staff working in a given country.

If the courier is traveling for the first time and must establish contact with an intelligence officer whom he does not know, he must make contact in person. When making contact with the courier, the intelligence officer must follow all the rules for establishing personal communication by appearance.

Organizing communications with a courier is, in principle, no different from communications between the station and agents in the target country.

Both the Center and the station must determine in advance the time, place and means of establishing contact with the courier. It is only necessary to additionally determine the call signals for the courier and the intelligence officer.

In some cases it may be necessary for the station chief to have a safe house ready to receive the courier. Then, upon arrival in the city, the courier appears at this address and establishes contact with the intelligence officer through this apartment.

Materials transported by couriers must be encrypted and disguised, keeping in mind the possibility of detention, search or arrest of the courier. Materials can be disguised in a variety of ways: hidden in a suitcase or a bag with a false bottom, sewn into clothes, writing a message on things in secret writing, embedded in food, household items, transmitted through microdots hidden in the courier's personal belongings, etc.

Transfer Points

A transfer point is an enterprise or establishment used by our intelligence organizations for the secret transfer of intelligence materials, operational, technical and financial means through an agent who is its owner or employee.

For a transfer point, a convenient cover can be a store, a snack bar, a library, a tobacco shop, a pharmacy, a beauty salon, a watch shop, a private clinic, a gas station, an auto repair shop, a garage and other enterprises and establishments that can be easily visited by a intelligence officer and an agent, explained by business considerations.

To transfer small-volume materials, you can use any of the listed places. To transfer large materials by volume, you can, for example, create a transmission point under the guise of a gas station, auto-assembly workshop, garage. It is convenient for the agent to go to such places put materials on the car, transfer them to the owner of the point, and the intelligence officer receives them afterwards.

Communication with agents through a transfer point is usually maintained in the following way: the agent appears under an appropriate pretext (this will depend on the nature of the point) to the agent running the transfer point and transfers intelligence material to him. Later, an intelligence officer comes to the transfer point, receives the agent's materials along with the purchases, and at the same time gives the owner of the point, if necessary, his instructions for transmission to the agent.

It is necessary that the behavior of the agent and intelligence officer who comes into contact with the holder of the point does not give the slightest reason for outsiders to suspect a tradecraft connection here. Each visit to a transfer point by an intelligence officer and agent must be externally justified.

This means that if the transfer point is organized under the guise of a bookstore, then when visiting it, the intelligence officer or agent should buy some book or at least inquire whether this or that book is on sale, so that for an outsider or counterintelligence agents had no doubts that this person entered the store specifically to buy a book. If there are other employees at the transfer point when the intelligence officer or agent arrives there, it is necessary that the intelligence officer and agent contact not only the owner of the transfer point, but also other employees if they also serve visitors.

The behavior of the intelligence officer and agent at the transfer point, their relationship with the owner of the point should not differ in any way from the behavior of other visitors.

Transfer points under the guise of such objects as a gas station, cafe, restaurant, store, library, which attract a large number of visitors, should be visited by the intelligence officer and agent not only on the days of exchange of intelligence materials, but also on other days in order to normalize their presence and to obscure their actions during an actual operation.

In order not to compromise the transfer point, the intelligence officer and agent must regularly visit other similar facilities: cafes, restaurants, shops, libraries, pharmacies, etc.

For the secret use of a transfer point, it is important to disguise the exchange of materials between the intelligence officer, the owner of the transfer point and the agent.

Masking methods should not be standard. They should be imaginative and should be determined by the nature and volume of materials, the characteristics of the transfer point and the environment in which the transfer takes place.

When transferring, for example, intelligence materials to the owner of a transfer point operating in a second hand bookstore, it is advisable to embed these materials in books. It is also necessary that the masking or embedding of materials in any things be done with the utmost care, since negligence in this can easily lead to undesirable consequences. The intelligence officer must constantly improve the techniques and methods of transmitting materials in order to make them as secret as possible, and train the agent in this.

When using a transfer point, it is necessary that its owner have specific signals for the intelligence officer and the agent to alert them about the presence of material and about danger. Typically, one transfer point is used to maintain communication with only one agent.

Clandestine Postal Addresses

Clandestine postal addresses are also used for intelligence communications. A secret address is an address secretly used by intelligence to receive mail in which intelligence and operational materials are hidden.

To create such addresses, verified, reliable agents are used from among individuals who enjoy a good reputation and are not suspected by counterintelligence. Such persons must be engaged in some business or have a circle of acquaintances and relatives, which would allow them to conduct regular, extensive correspondence.

In addition to persons who correspond with relatives or acquaintances, persons working as lawyers, businessmen, owners of hotels and various intermediary bureaus can be involved as holders of secret addresses.

In a number of eastern countries and in the provinces of some European countries, on the outskirts of cities and towns there are alleys and streets without house numbers. Residents of such areas, in order to guarantee the receipt of mail, provide the addresses of any nearby stores for correspondence and, with the consent of their owners, receive all correspondence to these addresses. The employee or owner of such a store may also be involved in intelligence work in order to use this address.

When sent to the holder of the secret address by mail or by telegraph, information should be kept secure. For these purposes, it is recommended to use secret codes and ciphers. The secrecy of messages must be so skillful that a secret letter or telegram looks no different from an ordinary letter or telegram. Conditional phrases entered during coding should not stand out from the text of the letter or telegram. The text should look completely natural and correspond to the usual correspondence of the address holder. If, for example, the owner of the address corresponds on commercial issues, then the clear text of the letter or telegram must also be disguise as commercial correspondence.

If parcels can be sent to the owner of the address or his correspondents, then a predetermined sequence of small items, for example, haberdashery and perfumes, can be used as a visual brevity code to transmit a certain signal, a short coded message, or pass on critical intelligence data.

The nature of the parcel must appropriate to the profession, the position of the addressee. It is most convenient to send intelligence materials using knowledge of microphotography. Always in a parcel or parcel It's good to camouflage the microdot.

If messages sent to a secret address are disguised as

personal correspondence of the owner of the address, it is necessary to examine his personal life, marital status, close relatives and address these issues in the letter.

You should pay attention to the language of the letter. The style of writing must correspond to the level of development, culture and literacy of both the holder of the secret address and its correspondent. If the holder of a secret address is the owner of a small shop, an uneducated and uncultured person, then it will be more natural to write letters in a style appropriate to his level with spelling errors, using popular expressions, since, when letters emanating from such a holder are written according to all grammatical rules, literary language, they can attract unwanted attention because they are out of place.

Letters should be written on a typewriter only when the correspondence is actually being conducted by the addressee, you need to use a typewriter whose font is similar to the typewriter of the holder of the secret address, but not the same. Moreover, the font can known to enemy counterintelligence as belonging to the intelligence organization of a socialist state.

Since the handwriting of intelligence officers from "legal" stations may be known to counterintelligence, they should not write letters in their own handwriting, but rather involve for this purpose persons whose handwriting is unknown to counterintelligence. At the same time, it is necessary to ensure that no fingerprints of the writer remain on the letters (when writing, printing and sending letters, you must work with gloves or place paper under it).

When sending secret letters, you cannot write a non-existent return address, since even a cursory check of such "addresses" leads to the establishment of the fact of deception and the recipient of the letter becomes an object of interest to the enemy counterintelligence organization. In exceptional cases, when there is no predetermined return address, it is better to indicate some hotel, boarding house, company, commercial office, and finally, a private address, but always an existing, valid address in the city from which the communication is sent.

It is not recommended to use the same type of paper and

envelopes for letters sent to different addresses. You cannot write letters to different recipients in the same handwriting.

It should be borne in mind that if the addressee is compromised, counterintelligence can reveal other addresses based on these signs.

When using a secret address for communication, it is necessary to know in detail the existing rules and norms for sending correspondence in a given country, the order of writing the address, the location of the stamp sticker, the cost of the stamp for intracity and out-of-town mailings, the rules of international correspondence, so as not to compromise the secret address and to ensure a reliable secret connection.

The agent who owns the secret address should not open letters intended for intelligence personnel. So that he can identify such letters, predetermined marks or icons are made on them.

Clandestine Telephone Services

A clandestine phone is a telephone secretly used by intelligence officers and agents to signal and transmit small coded operational and intelligence messages.

As the owner of a secret phone number, you also need to have a reliable and trusted agent. He may be recruited specifically for this purpose or be an agent who has lost his agent capabilities. The telephone can be installed in his apartment or at his place of work.

You can use a reliable person in secret as the keeper of a secret phone number. In this case, the proposal to use his phone can be motivated by reasons of a personal, business nature, or by a request to provide assistance to a certain organization whose activities he sympathizes with.

However, in all cases, it is necessary to ensure that the agent holding the secret telephone does not know the persons who use the telephone. To do this, the recruitment of such an agent should be entrusted to an intelligence officer who will not use this phone.

The keeper of a secret telephone number is subject to basically the same requirements as the keeper of a secret address; he must not be suspected by local counterintelligence agencies and must have a convenient cover.

The most convenient covers for these purposes are law firms and brokerage offices, doctors' offices, watch and car repair shops, and braiding studios, since their owners have a large clientele. If it's secret, the phone will be used correctly, then calls to similar enterprises and conversations between the intelligence officer and agent and the owner phones will not attract the attention of counterintelligence.

The correct use of such a telephone is as follows:

The intelligence officer and agent should not have open conversations with the owner of the phone about intelligence issues, but should use the code. The code must be drawn up taking into account the occupation of the keeper of the secret phone.

For example, if the owner of the secret phone number is a tailor, then the code should include as symbols the words usually used when talking with a tailor (fitting, lining, style, batting, suit, coat, date of fitting or receipt of the finished dress). Compliance with this requirement will prevent counterintelligence from unraveling the true nature of the conversation, even if it organizes eavesdropping.

The codes should be as simple as possible so that they can be easily remembered and not written down, and the messages transmitted should be brief. The agent and intelligence officer are prohibited from calling the owner of the telephone from work or from personal apartments. It is best to use a pay phone for this.

In a conversation with the owner of the telephone, the intelligence officer and agent should not mention their real names, but use a well-justified cover story and pseudonyms.
A conversation on intelligence issues with the owner of the telephone can only begin after the intelligence officer or agent,

having said the password, is convinced that the right person is on the line. The password and response should be completely natural and simple, so that people talking on the phone can accurately identify each other.

It is advisable to set certain hours for transmitting messages, and the owner of the telephone should be in his company or apartment at this time.

Clandestine Safe Houses

A safe house is an apartment secretly used by intelligence for secret operational purposes, for example, for meetings with agents, storage of operational equipment and materials, processing of intelligence materials, placement and operation of illegal radio equipment, etc. If necessary, safe apartments are specially equipped with hiding places for storage of materials, devices for secretly photographing materials, etc.

The apartments of proven agents who have lost their usual primary intelligence value, or the apartments of agents specially recruited for this purpose, are usually used as safe houses.

The loyalty of the agent (the owner of the apartment) and the suitability of the apartment itself are the main conditions that determine the possibility of its use. The safe house should be used for meetings, usually with only one agent. If an intelligence officer meets two or three agents in an apartment, this is a violation of the basic rules of secrecy.

CONCLUSION

Communication with agents must function clearly and reliably in any situation. This is especially important in the event of emergency circumstances, for example, the severance of diplomatic relations between the intelligence country and a socialist state, the outbreak of hostilities between countries in the imperialist camp, or a war between the imperialists and the countries of the socialist community.

If war breaks out, individual agents may be forced to go to other countries, some of them will be urgently called up for military service, and communication with loyal and valuable agents, which are extremely necessary for the intelligence of a socialist state, will be lost at this time.

To prevent this from happening, intelligence must carefully review the entire agent network in advance, select the most proven and valuable agents, work with which must continue under any conditions, and determine how communication with this agent will be carried out in the event of a crisis.

Under emergency conditions, decisions must be made on which of the agents will be transferred to contact illegal stations, who will maintain contact directly with the Center, and which agents will be consolidated into agent cells headed by a cell leader agent and how communication will be carried out with them.

With each promising agent, permanent and highly reliable channels of communication must be established that would make it possible to restore contact with him even in another country, if he is forced to travel and cannot inform the Centre about this travel. Therefore, it is imperative to agree with him how he should report his whereabouts.

When organizing communications, advance planning is essential in determining communication routes, methods and protocols for each agent in the network.

Types of communications that will be used for this; a method of communication with each agent included in one or another segment of this communication line. Safe houses and addresses, transfer points and hiding places must be prepared in advance, and the conditions for their use must be developed.

In the event that primary communication channels are compromised, secondary channels should be prepared to be activated in emergency situations.

In an operational environment, exceptional flexibility, maneuverability, ingenuity and courage will be required in finding such forms of communication with agents and methods of their use that would ensure its secrecy, reliability and efficiency.

www.ingramcontent.com/pod-product-compliance
Lightning Source LLC
Chambersburg PA
CBHW070128030426
42335CB00016B/2304